AF740127

# Lecture Notes in Geoinformation and Cartography

**Series editors**

William Cartwright, Melbourne, Australia
Georg Gartner, Wien, Austria
Liqiu Meng, München, Germany
Michael P. Peterson, Omaha, USA

*About the Series*

The Lecture Notes in Geoinformation and Cartography series provides a contemporary view of current research and development in Geoinformation and Cartography, including GIS and Geographic Information Science. Publications with associated electronic media examine areas of development and current technology. Editors from multiple continents, in association with national and international organizations and societies bring together the most comprehensive forum for Geoinformation and Cartography.

The scope of Lecture Notes in Geoinformation and Cartography spans the range of interdisciplinary topics in a variety of research and application fields. The type of material published traditionally includes:

- proceedings that are peer-reviewed and published in association with a conference;
- post-proceedings consisting of thoroughly revised final papers; and
- research monographs that may be based on individual research projects.

The Lecture Notes in Geoinformation and Cartography series also includes various other publications, including:

- tutorials or collections of lectures for advanced courses;
- contemporary surveys that offer an objective summary of a current topic of interest; and
- emerging areas of research directed at a broad community of practitioners.

More information about this series at http://www.springer.com/series/7418

Martin Breunig · Mulhim Al-Doori
Edgar Butwilowski · Paul V. Kuper
Joachim Benner · Karl Heinz Haefele
Editors

# 3D Geoinformation Science

The Selected Papers of the 3D GeoInfo 2014

Springer

*Editors*
Martin Breunig
Geodetic Institute
Karlsruhe Institute of Technology (KIT)
Karlsruhe
Germany

Mulhim Al-Doori
School of Engineering
American University in Dubai
Dubai
UAE

Edgar Butwilowski
Geodetic Institute
Karlsruhe Institute of Technology (KIT)
Karlsruhe
Germany

Paul V. Kuper
Geodetic Institute
Karlsruhe Institute of Technology (KIT)
Karlsruhe
Germany

Joachim Benner
Institute for Applied Computer Science
Karlsruhe Institute of Technology (KIT)
Karlsruhe
Germany

Karl Heinz Haefele
Institute for Applied Computer Science
Karlsruhe Institute of Technology (KIT)
Karlsruhe
Germany

ISSN 1863-2246 ISSN 1863-2351 (electronic)
Lecture Notes in Geoinformation and Cartography
ISBN 978-3-319-12180-2 ISBN 978-3-319-12181-9 (eBook)
DOI 10.1007/978-3-319-12181-9

Library of Congress Control Number: 2014957324

Springer Cham Heidelberg New York Dordrecht London
© Springer International Publishing Switzerland 2015
This work is subject to copyright. All rights are reserved by the Publisher, whether the whole or part of the material is concerned, specifically the rights of translation, reprinting, reuse of illustrations, recitation, broadcasting, reproduction on microfilms or in any other physical way, and transmission or information storage and retrieval, electronic adaptation, computer software, or by similar or dissimilar methodology now known or hereafter developed.
The use of general descriptive names, registered names, trademarks, service marks, etc. in this publication does not imply, even in the absence of a specific statement, that such names are exempt from the relevant protective laws and regulations and therefore free for general use.
The publisher, the authors and the editors are safe to assume that the advice and information in this book are believed to be true and accurate at the date of publication. Neither the publisher nor the authors or the editors give a warranty, express or implied, with respect to the material contained herein or for any errors or omissions that may have been made.

Printed on acid-free paper

Springer International Publishing AG Switzerland is part of Springer Science+Business Media (www.springer.com)

# Contents

# Improving the Consistency of Multi-LOD CityGML Datasets by Removing Redundancy

**Filip Biljecki, Hugo Ledoux and Jantien Stoter**

**Abstract** The CityGML standard enables the modelling of some topological relationships, and the representation in multiple levels of detail (LODs). However, both concepts are rarely utilised in reality. In this paper we investigate the linking of corresponding geometric features across multiple representations. We describe the possible topological cases, show how to detect these relationships, and how to store them explicitly. A software prototype has been implemented to detect matching features within and across LODs, and to automatically link them by establishing explicit topological relationships (with XLink). The experiments ran on our test datasets show a considerable number of matched geometries. Further, this method doubles as a lossless data compression method, considering that the storage footprint in the consolidated datasets has been reduced from their dissociated counterparts.

**Keywords** Multi-LOD · Topology · XLink · CityGML · Compression

## 1 Introduction

The OGC standard CityGML (Gröger and Plümer 2012; Open Geospatial Consortium 2012), and other 3D modelling standards such as COLLADA (ISO/TC 184: ISO/PAS 17506:2012; Sony Computer Entertainment Inc. 2008) and ISO's X3D

F. Biljecki (✉) · H. Ledoux · J. Stoter
Section GIS Technology, Delft University of Technology, Delft, The Netherlands
e-mail: f.biljecki@tudelft.nl

H. Ledoux
e-mail: h.ledoux@tudelft.nl

J. Stoter
Kadaster, Product and Process Innovation, Apeldoorn, The Netherlands
e-mail: j.e.stoter@tudelft.nl

J. Stoter
Geonovum, Amersfoort, The Netherlands

© Springer International Publishing Switzerland 2015
M. Breunig et al. (eds.), *3D Geoinformation Science*,
Lecture Notes in Geoinformation and Cartography,
DOI 10.1007/978-3-319-12181-9_1

(ISO/IEC 19775-1:2013) allow the storage of multiple level of detail (LOD) representations of a model, in order to facilitate the multi-scale use of the models and to improve the computational efficiency of spatial operations (Biljecki et al. 2014a).

Although 3D GIS datasets may contain multiple LODs, multi-LOD datasets are almost non-existent in practice, and they are seldom linked beyond an administrative link between object identifiers (i.e. they share the same building ID) (Biljecki et al. 2013). In our opinion, this situation is caused by the following deficiencies: (1) lack of consistency, i.e. there is redundancy in the acquisition–modelling–storage process; (2) when using multi-LOD datasets, it is not always clear when and how to switch between LODs as it is the case in computer graphics; and (3) 3D generalisation specifications and implementations are not fully developed, limiting the generation of LODs other than the one being primarily constructed from the acquired dataset.

In this paper we focus on the first shortcoming by investigating the possible improvements in the consistency and storage of multi-LOD datasets, with a theory that is applicable also to single-LOD representations. It is our experience that in practice, besides exemplary models, single-LOD datasets do not contain the explicit representation of topological relationships, hence developing a joint method that is beneficial for both possibilities is important. We observe and take advantage of the practical fact that many of the stored geometries (primarily polygons) in 3D datasets are geometrically equal both within a single LOD and across multiple LODs. By determining the topological relationships between such reoccurring geometries and storing them explicitly, the consistency of 3D models can be increased, as we show in this paper. However, while developing the method, we have realised that in practice most of the geometries that are reoccurring are not identical and cannot be readily matched. Therefore, we have investigated other cases and covered them as well.

This paper is focused towards CityGML and its LOD concept, however, most of the developed work is applicable to other formats. Our work consists of the following contributions: (1) we have investigated and described several cases of reoccurring geometries and introduce a terminology to distinguish them; (2) we have developed robust algorithms that efficiently index the geometries in CityGML datasets and that take advantage of the geometries that reoccur by explicitly storing their topological relationships; (3) we have developed a software prototype that analyses CityGML data and automatically computes explicit topological links between matching geometric features through the XML Linking Language (XLink) mechanism; and (4) we show with experiments that a considerable subset of data can be matched. We have tested the method on a synthetic dataset that contains buildings in LOD1, LOD2, and two variants of LOD3.

Because matched geometries are stored only once, the consolidated dataset is compressed without loss of information.

In Sect. 2 we explain the advantages of establishing explicit topological representations between geometric features. In Sect. 3 we describe possible topological cases, introduce our terminology, and present the algorithms that index, match, and consolidate the geometry of 3D datasets, primarily CityGML. The consolidation of the data is a technical challenge because it is done on a hierarchical data structure, and there might not be an optimal algorithm that is suited for all CityGML datasets.

The implementation and the results, presented in Sect. 4, show that after linking a higher degree of the consistency of the data is achieved, contributing to an efficient storage and maintenance. For instance, if the geometry of a feature is altered in one LOD, thanks to the established explicit topological representations this change may propagate through other LODs.

## 2 Background and Related Work

Consistency of 3D city models is an important topic in GIS, in which topology plays a prominent role (Gröger and Plümer 2009; Ledoux and Meijers 2011). Current research efforts focus on the relationships of features within the same representation, e.g. the validation of solids (Ledoux 2013), and making use of topological data structuring to improve rendering performance of 3D city models on mobile devices (Ellul and Altenbuchner 2014). To the extent of our knowledge, there is no related work to detect and link geometric features across multiple representations.

While this paper generally describes a way how to increase the consistency and to compute topological relationships in a model, it is focused on the maintenance and storage of (3D) GIS datasets, which is a topical subject in academia and industry (Aringer and Roschlaub 2014; Steinhage et al. 2010; Stoter et al. 2011). Updates of models often introduce errors (Gröger and Plümer 2012), so increasing consistency is one of the prerequisites for an efficient workflow.

In this section we describe the redundancy and benefits of an established topology with respect to the scalability of the models: for single representations (single-LOD datasets), and for multiple representations (multi-LOD datasets).

### *2.1 Single-LOD Datasets*

Research that has been done in this topic is focused on the relationship of real-world features within the same representation (Gröger and Plümer 2005, 2011). For instance, the topology of two coinciding polygons, such as a wall shared by two buildings. The consistency that is achieved by establishing explicit topological relationships in practice simplifies the maintenance of the data and reduces the redundancy in the storage.

Figure 1 shows an example of the benefit with respect to the maintenance of a 3D model. The left model (Fig. 1a) shows a building with a wall that contains a window. The polygon representing the wall is shown in red, and it contains a hole (inner ring), which is filled by another polygon representing the window. The interior ring of the wall polygon corresponds to the exterior ring of the polygon representing the window. In a model without established explicit topological relationships, the two features are not linked in any way. When the geometry of a part of the object is updated, e.g. the window is enlarged, the change does not affect

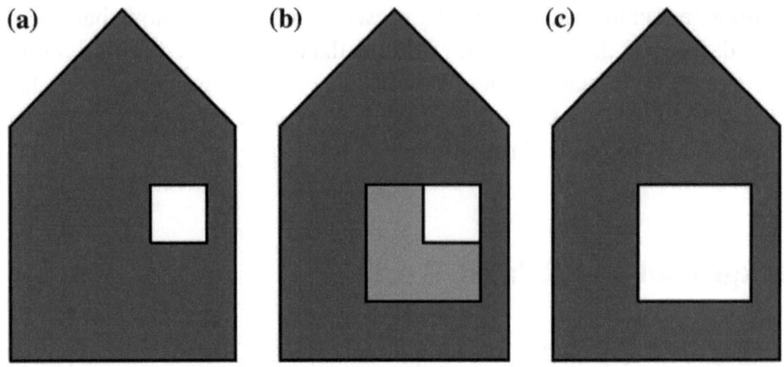


**Fig. 1** The determined explicit topological relationships in a dataset has a significant benefit to its maintenance. This example shows the benefit on a wall with a window that is being enlarged. **a** Model of a building before the update (enlargement of the window). **b** Updated model without links resulting in inconsistency. **c** Updated model with established links (desired case)

the related geometry (i.e. hole of the wall), leading to redundancy in the process (see Fig. 1b). In a model with established topological representations, the change properly propagates to the related features (see Fig. 1c for the desirable outcome).

## *2.2 Multi-LOD Datasets*

On top of the redundancy in a single representation, that multiplies with each new representation in a multi-LOD dataset, there is also additional redundancy. For instance, should a feature be changed, in practice the update must be done manually for each representation. Further, because of the increasing complexity of the models, the size of the datasets substantially increases with the increase of the LOD, making the storage less feasible. The surge in the size is not only caused by the growth of the amount of details, but also because of the redundancy that could be removed, as we show later in the paper.

Therefore, despite the fact that this option is available in CityGML, models are usually derived in a single LOD representation. While multi-LOD datasets are rare, when they are available they are usually produced by generalisation from finer LODs (e.g. see Akmalia et al. 2014; Baig and Abdul-Rahman 2013; Zhao et al. 2012), for instance, as a bounding box of an LOD2 (El-Mekawy and Östman 2011), or of an LOD3 including features such as antennas on roofs (Mao et al. 2012). This is beneficial for this research, since it results in datasets where many of the geometries are preserved, and are identical in more than one representation. For instance, the ground surface of a building (i.e. Ground Surface) is usually identical in all representations.

Detecting and linking such occurrences would be a first step towards complementing the discussed practical shortcomings.

# 3 Methodology

We have developed a method that searches for matching geometries in the datasets and links them. Figure 2 shows the desirable outcome of the algorithms with an example of three LODs where some of the geometries are reoccurring and are consequently linked.

While examining multi-LOD datasets we have realised that there are different cases of corresponding geometry, not only polygons that are identical and that can be directly referenced. For instance, polygons that share the exterior ring, but their interior is different (common in CityGML LOD3 where openings are allowed, see Fig. 2). Further, specific cases such as two equal polygons whose starting point is different should also be handled.

## *3.1 Terminology*

In this paper we focus on the two geometric feature types: polygons and linear rings. We consider two or more geometric primitives *identical* if they are topologically and geometrically equivalent, i.e. they can be readily linked and re-used. The geometric representations of the ground plane of a building in two LODs are usually identical. Two or more primitives are *partially identical* if they are not identical and if their relation has one or more of the following properties which prevent them to be identical:

- The orientation of their vertices is different, i.e. their normals are reversed. For instance, two buildings share the same wall.

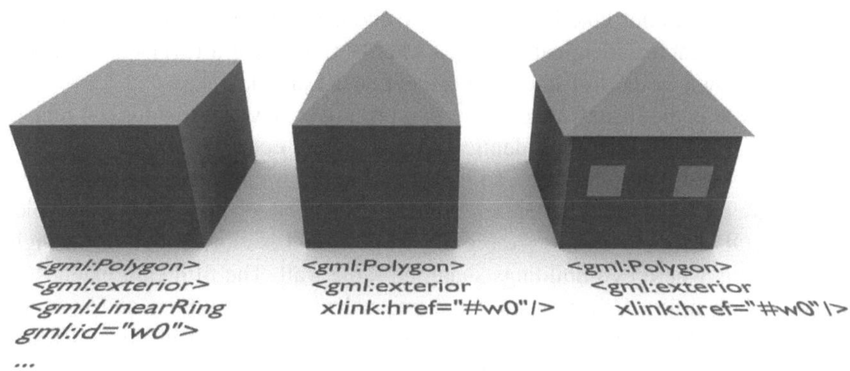


**Fig. 2** The rationale of the method. If two or more geometries are found to correspond, links are created. In this case two polygons (walls) are identical in two LODs, however, in one LOD the polygon has a hole, hence, only their exterior rings are linked. How and where to establish the links while balancing the maximisation of links and the topological structure is the main concern of the consolidation process

- They constitute different aspects of their parent primitive, e.g. two linear rings correspond, but one forms the exterior ring of its parent polygon, while the other forms the interior ring to describe a hole. A prominent example of this case is a wall with a hole that represents an opening—window or door which is stored separately.
- The number of points in their rings is not equal while the shape and location are identical. This is caused by *redundant* points $p_i$, where $p_{i-1}$, $p_i$, $p_{i+1}$ are collinear. The removal of such points would not compromise the shape and location of the polygon.
- The starting point in the linear ring is different. This discrepancy might be easily detected and corrected *on-the-fly*, hence it will not be particularly emphasised in the continuation of the paper.

Two geometries *match* if they are either identical or partially identical. When a match of two or more geometries is found, one is selected as the *resource*, and the rest are *linked* to it.

## 3.2 Topological Relationships

In this section we show the possible cases of the *matching* geometry, i.e. topological relationships between polygons, and their constituting components—exterior and interior ring(s). We have investigated the possible cases, and their occurrences in real-world datasets, which we show in Table 1. The sign = denotes identical primitives,—partially identical, ≠ non-matching, and ∅ denotes no geometry. The matched primitives in each case are shown in red. In the example column, when two objects are separated, the example refers to the multi-LOD case.

While more "permutations" are possible, they are not valid according to GML, hence the list does not take into account invalid cases. One such example are two exterior rings that are identical, but their interior rings have a reversed orientation (which is shown in the last line as an exceptional example).

The case 0 is the (usual) case where two primitives have no topological relation in the context of this research. Identical features rarely occur within the same LOD (in contrast to partially identical), so the cases 1 and 2 where the geometries of two features are identical are often present in multi-LOD datasets. Case 3 is also typical in multi-LOD datasets (again, see Fig. 2), while case 4 is more common in single-LOD datasets where two buildings share the same wall. The fifth case extends the previous with holes. Cases 6 and 7 are unusual in the real-world, and case 8 is similar to the fifth case in the occasion when the polygon is not identical. Cases 9 and 10 are cases of interchangeable roles of the exterior and interior rings, and

**Table 1** Cases of topological relationships of rings and polygons

| Case | Ring Exterior | Ring Interior | Polygon | Graphical explanation | Real-world example |
|---|---|---|---|---|---|
| 0 | ≠ | ≠ | ≠ |  | 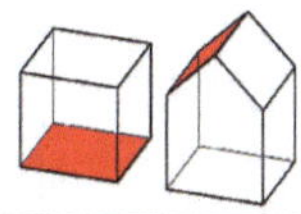 |
| 1 | = | ∅ | = | 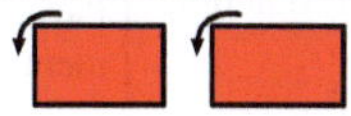 | 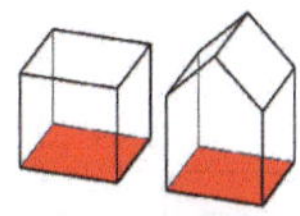 |
| 2 | = | = | = | 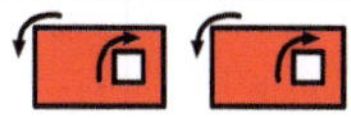 | 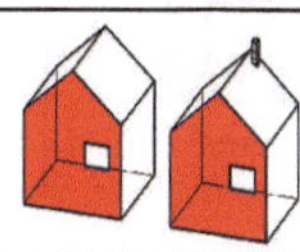 |
| 3 | = | ≠ | ≠ | 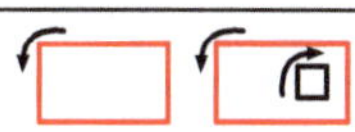 | 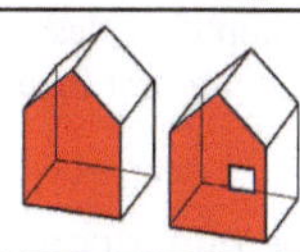 |
| 4 | – | ∅ | – | 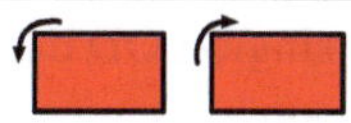 | 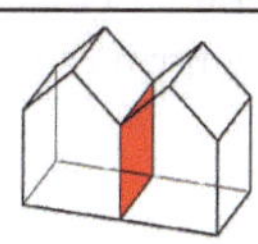 |
| 5 | – | – | – |  | 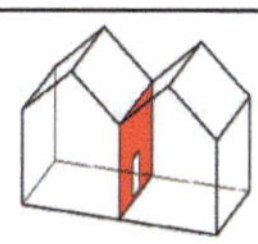 |
| 6 | – | ≠ | ≠ | 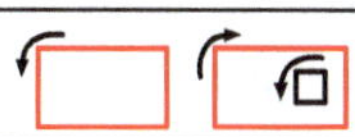 | Unidentified |
| 7 | ≠ | = | ≠ | 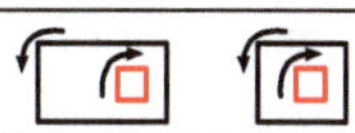 | Unidentified |
| 8 | ≠ | – | ≠ |  | 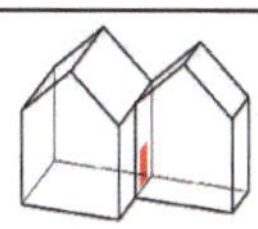 |
| 9 | ⟵ = ⟶ | | ≠ |  | Unidentified |
| 10 | ⟵ – ⟶ | | ≠ |  |  |
| Invalid | = | – | ? | 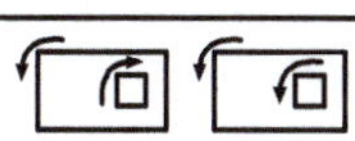 | Not possible |

The curved arrows denote the ringâ€™s orientation, while the long horizontal arrows indicate that there is a relation between one polygonâ€™s exterior to another polygonâ€™s interior

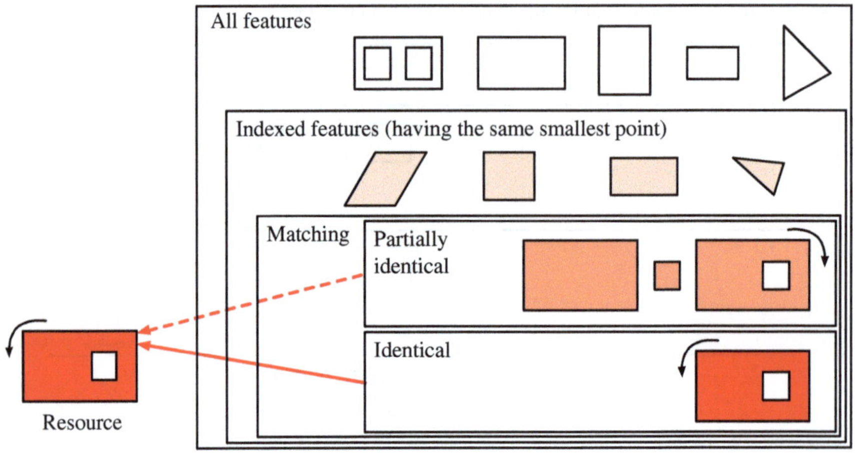


**Fig. 3** Simplified classification of the relationships and workflow of the method. The primitives in the dataset are first indexed, and then tested for matches, of which there are different categories

might be rather considered as extended cases. Case 9 is uncommon, and case 10 is usually occurring in finer single-LOD datasets as the relation between a wall and an opening.

### *3.3 Overview of the Method*

The workflow of the method to match regards both matching primitives within the same LOD and across multiple LODs:

1. Indexing. Index all points, linear rings, and polygons in all LODs for efficiency (Sect. 3.4).
2. Matching. Detect matching geometries and flag them. Because of different topological relations, which have been introduced in the previous section, the detection of the matching geometries is done in multiple phases (Sect. 3.5).
3. Consolidation. Analyse the matched geometry and remove redundant data by replacing them with a link to one other matching representation resource, with modifications if necessary (Sect. 3.6).

Figure 3 shows the simplified workflow of the method and the relation between the features when searching for matches. Because all features are indexed, the searching algorithm has a considerably reduced subset of potential matches. After ruling out non-matching features in the indexed subset, the algorithms detect the matched features and classify them according to the cases presented in the previous section.

The developed algorithm and the implemented software prototype are focused towards ideal cases where the vertices of the geometry are identical across LODs and where the geometries fully correspond. This is useful for datasets produced with generalisation, however, when used on datasets with a different lineage it might not produce results to the same extent. This could be solved by introducing the snapping of the points according to a tolerance threshold, and more sophisticated matching of similar geometries. The automatic matching of the representations that are acquired with different techniques would require employing more advanced algorithms such as (Arkin et al. 1991) and (Zheng et al. 2014), extending related work done in cartography (e.g. Dilo et al. 2009; Zhang et al. 2014) to 3D GIS, and would probably result in a lossy compression (partial data discarding).

## *3.4 Algorithm for Indexing the Geometry*

In order to make the search for the matching geometry more efficient and enable the consolidation of larger datasets, the polygons and their rings are first indexed. We have decided to build an index where each polygon's ring is indexed according to its smallest point. The smallest point is the point with the smallest coordinate value, i.e. the one that is closest to $(-\infty, -\infty, -\infty)$. Each valid ring has one such point, hence, it can serve for the purpose of indexing. This point should not be confused with the starting point of the ring, which is not relevant here.

This step considerably reduces the search time for matches since in practice only a few non-matching polygons share the same smallest point. As an example, Fig. 4 shows two different polygons that share the same smallest point. The comparison of such polygons for discarding the different geometries is described in the next step.

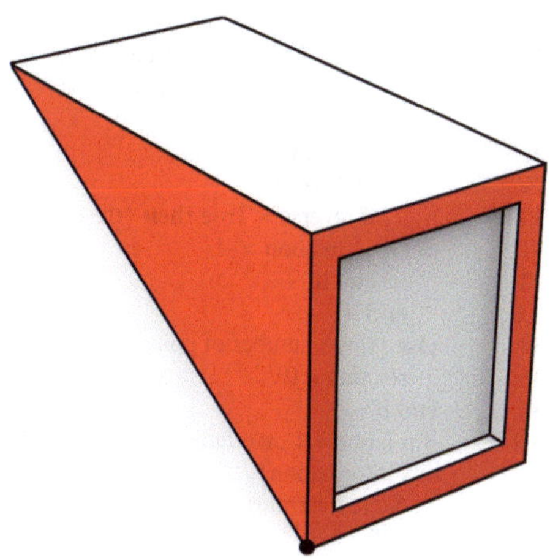

**Fig. 4** Multiple different polygons (shown here in *red*) may have the same smallest point (shown in *black*). In this case the two polygons are part of a dormer of a building

### 3.5 Algorithm for the Detection of Matching Geometries

After indexing, the rings (both exterior and interior) are queried for their relations. The algorithm first removes vertices that are redundant (i.e. being collinear with its preceding and succeeding points). The algorithm is given in Algorithm 1.

---

**Algorithm 1** Algorithm for the detection of matching geometries.

**Input:** Indexed features, where each ring has its smallest point indexed (Sec. 3.4)
**Output:** Topological relationships between the geometries

1: Iterate all linear rings and remove all redundant points
2: **for** each ring $r_j$ of each polygon **do**
3:   **for** each ring $r_k$ that shares the same smallest point **do**
4:     relation = 1 {Assume identicalness until proven otherwise}
5:     **if** $r_j$ and $r_k$ are identical features **then** {Avoid comparing the ring to itself}
6:       Go to 3
7:     **end if**
8:     **if** the relation between the two rings has already been checked **then**
9:       Go to 3
10:     **end if**
11:     **if** the number of points of $r_j$ and $r_k$ are equal **then**
12:       Reinstate the points{i.e. reorder them so the first point coincides}
13:       **while** relation = 1 **do**
14:         **for** each point $p^i_j$ in $r_j$ **do** {Start from the second point because the first one is identical in any case (index)}
15:           **if** $p^i_j \neq p^i_k$ **then** {Points not identical, so there is no match}
16:             relation = 0
17:           **end if**
18:         **end for**
19:       **end while**
20:       **if** relation = 0 **then** {If the match has not been found, try for the other orientation}
21:         reverse = True {Assume that the linear rings are inverted until proven otherwise}
22:         n = number of points
23:         **while** reverse = True **do**
24:           **for** each point $p^i_j$ in $r_j$ **do** {Start from the second point because the first one is equal in any case (index)}
25:             **if** $p^i_j \neq p^{n-i}_k$ **then** {Points not identical when the rings are reversed}
26:               reverse = False
27:             **end if**
28:           **end for**
29:         **end while**
30:         **if** reverse = True **then** {the relation between the rings is reversed}
31:           relation = -1
32:         **end if**
33:       **end if**
34:     **else** {If the number of points is different, the features are not identical}
35:       relation = 0
36:     **end if**
37:     **if** relation = 1 **or** relation = -1 **then** {If the rings are identical or partially identical, store this information}
38:       Store the information about the relation $r_j \leftrightarrow r_k$
39:     **end if**
40:   **end for**
41: **end for**

---

### 3.6 Algorithm for the Consolidation of the Data

After detecting matching geometries, the last phase of the method involves consolidating the data (linking), i.e. analysing the relationships and determining the level of the relationship that can be linked. A straightforward solution would be to directly link the matching rings. However, because of the different cases and hierarchies, this phase is not forthright, and it can be solved in multiple ways. For instance, if two rings that form the exterior of two polygons match, this does not necessarily mean that the polygons can be matched right away, since the interior may be different (e.g. see case 3 in Table 1 ). Further, one of these rings may be an interior of another polygon, that is further related to another polygon in another way. Therefore, maximising the number of links that are established is the main concern when designing such algorithm, and cannot be solved simply by determining the frequency of the occurrences and selecting the topmost ring. Further, the development of the algorithm is associated with the content of a targeted dataset, as an algorithm might not be equally beneficial when employed for consolidating two different datasets. This problem is related to the field of data compression (Huffman 1952; Salomon et al. 2007).

We have designed a top-down approach that first iterates polygons with holes, comparing the matched rings, and builds a hierarchy of features, continuing to polygons without holes. This is particularly beneficial for cases 1, 2, 3 and 10, which are the most common. The algorithm is given in Algorithm 2.

## 4 Implementation and Results

### 4.1 Test Data

Because multi-LOD datasets are rare in practice, there is a difficulty to find input material for the testing of the prototype. The publicly available CityGML datasets that contain more than one LOD representation are limited to one or a few buildings (e.g. Häfele 2011).

We have used a dataset in multiple LODs that was automatically generated from a parametrised description by the engine "Random3Dcity", which was presented in Biljecki et al. (2014b). The dataset contains 100 buildings. Some of the LODs are represented in two ways, as a <gml:Solid> or as semantically structured surfaces (<gml:boundedBy>). This is done in order to extend the experiments by comparing the different variations of the models that are valid (Benner et al. 2013; Löwner et al. 2013). Further, two variants of LOD3 are available, in order to take into account the different levels of complexity that may occur in the same LOD range (see Biljecki et al. 2014a for further details). The dataset contains six representations of each building, that are shown in Fig. 5.

**Algorithm 2** Algorithms for consolidating the data.

**Input:** Topological relationships, obtained from the previous algorithm in Sec. 3.5
**Output:** Consolidated dataset with established links

```
1: for each polygon P0 that contains the interior and has at least one ring matching to a ring of at
   least one other polygon Pi do {Both exterior and interior rings are taken into account}
2:   if P0 and Pi were already matched then
3:     Go to 1
4:   end if
5:   if Pi comprises only polygons with holes and are identical to P0 then
6:     Establish P0 as the resource, and link Pi
7:     Remove the primitives from further consideration
8:   end if
9:   if the exterior of Pi is identical to P0 then
10:    Establish the exterior of P0 as the resource, and link Pi
11:    Remove the primitives from further consideration
12:  end if
13:  if Pi is partially identical to P0 then
14:    Establish the exterior of P0 as the resource with necessary modifications, and link Pi
15:    Remove the primitives from further consideration
16:  end if
17: end for
18: for each polygon P0 that does not contain the interior and has at least one ring matching with
    a ring of at least one other polygon Pi do
19:  if P0 and Pi were already matched then
20:    Go to 18
21:  end if
22:  Repeat the steps above without the operations on the interior
23: end for
24: Store the consolidated dataset with determined topological relationships
```

LOD1s The standard LOD1 block model, represented as a <gml:Solid>. The top of the block model represents the height of the building at the eaves. The footprint represents the *real* footprint of the building (*cf.* cadastral records).

LOD2s A model with simple standardised roof shapes. The footprint is the same as in LOD1. It is also stored as a solid.

LOD2b The semantically enriched boundary representation from which the previous model was generated.

LOD3s A solid obtained from the architecturarly detailed model. In comparison to the LOD2, it includes dormers and other objects that contribute to the internal volume of the building.

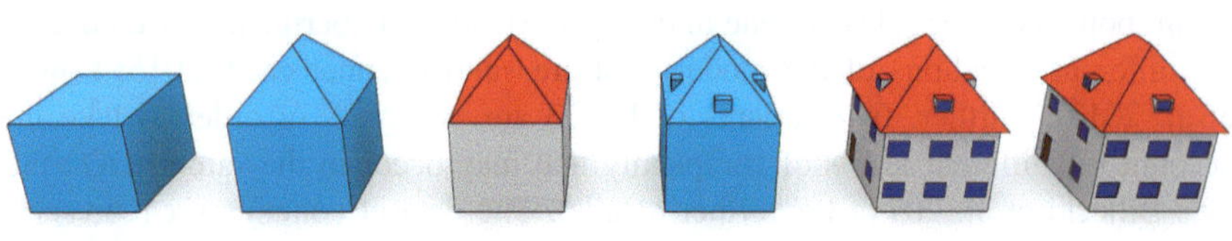

**Fig. 5** Visualisation of the six representations that are available in the test data. Their order from the left is the same as in the description in the text

LOD3b A detailed model with openings (doors and windows). From the geometric perspective, the major difference with respect to the previous representations is that the polygons contain <gml:interior>.

LOD3+ A very detailed variant of LOD3 that contains smaller details such as embrasures of windows and doors (see Fig. 4 for example). Such models are rare in CityGML. This representation is important because it presents a considerable increase in the storage footprint comparing to the coarser representations, e.g. roughly twice the size of LOD3b.

Since CityGML does not support the simultaneous representation of more than the five standard LODs, the datasets have been stored in separate files.

## 4.2 Implementation

The implementation was done in Python. The XML Linking Language (XLink) mechanism was used to realise the links between the features. It allows elements to be inserted into Extensible Markup Language (XML) documents for creating and describing links between resources (XML Core Working Group 2010).

XLink has been used already in research projects that employ CityGML (Iwaszczuk and Stilla 2010; Kolbe 2009; Stadler et al. 2009). It is also mentioned in the GML standard (Lake et al. 2004; Open Geospatial Consortium 2012) and in the CityGML standard (Open Geospatial Consortium 2012) as the preferred way of explicitly storing the topological relationships. In CityGML, it is primarily used for referencing mutual geometries of two objects (e.g. a building and a building part), and for the re-use of the geometry in the semantically enriched boundary representation for the construction of solids (Gröger and Coors 2011).

The example below shows the code for a resource and a link. When two or more matching geometries are detected, the prototype adds an identifier to the gml:id attribute of the resource, i.e. the preserved single instance of the multiple matches. For the unambiguous identification of resources and links, the Universally Unique Identifier (UUID) standard has been used (ISO/IEC 2008).

```
<gml:surfaceMember
  <gml:Polygon gml:id="0127875b-a2a8-498e-8024-770fd661aef6">
    <gml:exterior>
      <gml:LinearRing>
        <gml:posList>
          173469.0 526442.0 0.0
          173469.0 526446.64 0.0
          173478.81 526446.64 0.0
          173478.81 526442.0 0.0
          173469.0 526442.0 0.0
        </gml:posList>
      </gml:LinearRing>
    </gml:exterior>
  </gml:Polygon>
</gml:surfaceMember>
```

Afterwards, the contents of each matching geometry is removed and a link is added pointing to the resource which contains the geometry:

```
<gml:surfaceMember xlink:href="#0127875b-a2a8-498e-8024-770fd661aef6"/>
```

Geometries that have opposite normals (reversed ring orientation) can be referenced with <gml:OrientableSurface>. For instance, the reversed match of the polygon above may be stored as:

```
<gml:OrientableSurface orientation="-">
  <gml:baseSurface xlink:href="#0127875b-a2a8-498e-8024-770fd661aef6"/>
</gml:OrientableSurface>
```

For reversing the linear rings, <gml:OrientableCurve> can be used. These examples not only show the achieved consistency, but also the decrease in the storage footprint.

## *4.3 Results*

The implemented software prototype was run on the test dataset that contains diverse cases of matching geometries. After running, the results show considerable improvements in the consistency of the dataset. A significant number of polygons was found to match, both within single-LOD and multi-LOD data. Further, examining the results in details shown that some of the rings reoccured as many as 9 times, which reinforces the importance of linking such geometries.

The total number of polygons in the datasets is 13,021 (of which 8 % has interior) with 17,029 rings. Indexing reduced the amount of queries by 99.88 %. Most of the points in the index referred to less than 10 rings (i.e. the number of rings that share the same smallest point), and within each comparison, an average of 42 % of rings were found to match. The geometries were linked, and the consolidated data has been stored.

A useful insight in the established links is that all interior rings in the dataset are found to be matched to an exterior ring of another polygon. This is due to openings such as windows and doors, and closure surfaces.

Obviously, the obtained results such as the number of matched primitives (88 %), strongly depend on the used dataset, however, they give a good impression of the quantity of reoccurring geometries, for which there is no reason to store them more than once.

While the consolidation algorithm provides a good balance between simplicity and the end result (amount of consolidated data), its computational complexity is exponential, rendering it less feasible for larger datasets.

From the storage perspective, after the consolidation the size of the dataset was reduced by 20 %. Due to the highly repetitive structure of an XML schema such as

CityGML, we do not expect that the consolidation of recurring geometries can further contribute to the reduction of the storage footprint. Further data compression of multi-LOD datasets should be sought in methods such as XML clustering (Dalamagas et al. 2004).

## 5 Conclusions and Future Work

In this paper we have presented a method to analyse multi-LOD CityGML datasets by detecting matching geometries, and to automatically adapt them by storing the explicit representations of their topological relationships. This enhancement considerably improves the consistency of the model, leading to more efficient maintenance and storage. The software prototype that we have implemented shows significant improvements in that respect.

As a foundation of the work, we have developed a theoretical framework that describes cases of topological relationships occurring in reality, and we investigate more closely the XLink mechanism that provides the explicit modelling of some topological relationships in (City)GML. The work can be used for both within a representation and across multiple representations, doubling its purpose. Since our implementation is automatic, we hope that these improvements will contribute to the increased creation of multi-LOD datasets and their deeper integration.

While the tests involved only buildings, the presented algorithms are intended to work for other thematic classes as well. However, because of the lack of such datasets, testing possibilities are limited, and had to be done on a synthetic dataset produced by our random engine (Biljecki et al. 2014b).

The matching algorithm is suited for ideal cases such as generalised datasets which contain identical primitives across LODs, and may fall short in datasets constructed with different acquisition techniques. Improvements in this direction are one of the aims that we plan for future work. Further, we plan to improve the algorithm that consolidates the data and to offer a few alternatives with different advantages and suitability.

**Acknowledgements** We would like to thank our colleague Ken Arroyo Ohori for his suggestions, and the anonymous reviewers for their thorough comments which have helped us to improve the manuscript. This research is supported by the Dutch Technology Foundation STW, which is part of the Netherlands Organisation for Scientific Research (NWO), and which is partly funded by the Ministry of Economic Affairs. (Project code: 11300)

## References

Akmalia R, Setan H, Majid Z, Suwardhi D, Chong A (2014) TLS for generating multi-LOD of 3D building model. In: IOP conference series: earth and environmental science. Proceedings of the 8th international symposium of the digital earth (ISDE8) vol. 18, pp 1–8

Aringer K, Roschlaub R (2014) Bavarian 3D building model and update concept based on LiDAR, image matching and cadastre information. In: Innovations in 3D geo-information sciences. Springer, Berlin, pp 143–157

Arkin EM, Chew LP, Huttenlocher DP, Kedem K, Mitchell JSB (1991) An efficiently computable metric for comparing polygonal shapes. IEEE Trans Pattern Anal Mach Intell 13(3):209–216

Baig SU, Abdul-Rahman A (2013) Generalization of buildings within the framework of CityGML. Geo-spatial Inf Sci 16(4):247–255

Benner J, Geiger A, Gröger G, Häfele KH, Löwner MO (2013) Enhanced LOD concepts for virtual 3D city models. In: Isikdag U (ed.) ISPRS annals of the photogrammetry, remote sensing and spatial information sciences. Proceedings of the ISPRS 8th 3D geoinfo conference and WG II/2 workshop, Istanbul, Turkey, pp 51–61

Biljecki F, Zhao J, Stoter J, Ledoux H (2013) Revisiting the concept of level of detail in 3D city modelling. In: Isikdag U (ed.) ISPRS Annals of the photogrammetry, remote sensing and spatial information sciences. Proceedings of the ISPRS 8th 3D GeoInfo conference and WG II/ 2 workshop, Istanbul, Turkey, pp 63–74

Biljecki F, Ledoux H, Stoter J, Zhao J (2014a) Formalisation of the level of detail in 3D city modelling. Comput Environ Urban Syst 48:1–15

Biljecki F, Ledoux H, Stoter, J (2014b) Error propagation in the computation of volumes in 3D city models with the monte carlo method. In: ISPRS annals of the photogrammetry, remote sensing and spatial information sciences. Proceedings of the ISPRS/IGU joint international conference on geospatial theory, processing, modelling and applications, Toronto, Canada

Dalamagas T, Cheng T, Winkel KJ, Sellis T (2004) Clustering XML documents by structure. In: Panayiotopoulos T, Vouros GA (eds) Lecture notes in computer science, methods and applications of artificial intelligence. Springer, Berlin, pp 112–121

Dilo A, van Oosterom P, Hofman A (2009) Constrained tGAP for generalization between scales: The case of Dutch topographic data. Comput Environ Urban Syst 33(5):388–402

Ellul C, Altenbuchner J (2014) Investigating approaches to improving rendering performance of 3D city models on mobile devices. Geo-spatial Inf Sci 17(2):73–84

El-Mekawy M, Östman A, Shahzad K (2011) Towards interoperating CityGML and IFC building models: a unified model based approach. In: Kolbe TH, König, G, Nagel C (eds) Proceedings of the 5th international 3D geoinfo conference. Advances in 3D geo-information sciences. Lecture notes in geoinformation and cartography, Springer, pp. 73–93

Gröger G, Coors V (2011) Modeling guide for 3D objects. Tech. rep., SIG3D

Gröger G, Plümer L (2005) How to get 3-D for the price of 2-D? Topology and consistency of 3-D urban GIS. GeoInformatica 9(2):139–158

Gröger G, Plümer L (2009) How to achieve consistency for 3D city models. GeoInformatica 15 (1):137–165

Gröger G, Plümer L (2011) Topology of surfaces modelling bridges and tunnels in 3D-GIS. Comput Environ Urban Syst 35(3):208–216

Gröger G, Plümer L (2012a) CityGML—Interoperable semantic 3D city models. ISPRS J Photogrammetry Remote Sens 71:12–33

Gröger G, Plümer L (2012b) Transaction rules for updating surfaces in 3D GIS. ISPRS J Photogrammetry Remote Sens 69:134–145

Häfele KH (2011) CityGML model of the FJK-Haus. Institut für Angewandte Informatik (IAI), Karlsruher Institut für Technologie

Huffman D (1952) A method for the construction of minimum-redundancy codes. Proc IRE 40 (9):1098–1101

ISO/TC 184: ISO/PAS 17506:2012 (2012) Industrial automation systems and integration—COLLADA digital asset schema specification for 3D visualization of industrial data

ISO: ISO/IEC 19775-1:2013 (2013) Information technology—Computer graphics, image processing and environmental data representation–Extensible 3D (X3D)—Part 1: architecture and base components (2013)

ISO: ISO/IEC 9834-8:2008(E) (2008) Information technology—open systems interconnection—procedures for the operation of OSI registration authorities: generation and registration of universally unique identifiers (UUIDs) and their use as ASN.1 object identifier components

Iwaszczuk D, Stilla U (2010) A concept for assignment of textures to partially occluded faces of 3D city models stored in CityGML. In: Kolbe TH, König G, Nagel C (eds) International archives of photogrammetry, remote sensing and spatial information sciences. Proceedings of the 5th international 3D geoinfo conference, Berlin, Germany, pp 57–62

Kolbe TH (2009) Representing and exchanging 3D city models with CityGML. In: Zlatanova S, Jiyeong L (eds) 3D geo-information sciences. Springer, Berlin, pp 15–31

Lake R, Burggraf D, Trninić M, Rae L (2004) Geography mark-up language (GML), foundation for the geo-web. Wiley, UK, pp 1–408

Ledoux H (2013) On the validation of solids represented with the international standards for geographic information. Comput-Aided Civ Infrastruct Eng 28(9):693–706

Ledoux H, Meijers M (2011) Topologically consistent 3D city models obtained by extrusion. Int J Geog Inf Sci 25(4):557–574

Löwner M-O, Benner J, Gröger G, Häfele KH (2013) New concepts for structuring 3D city models —an extended level of detail concept for CityGML buildings. In: Murgante B, Misra S, Carlini M, Torre CM, Nguyen HQ, Taniar D, Apduhan BO, Gervasi O (eds) Lecture Notes in Computer Science. Proceedings of the 13th international conference computational science and its applications–ICCSA 2013, Ho Chi Minh City, Vietnam, pp 466–480

Mao B, Harrie L, Ban Y (2012) Detection and typification of linear structures for dynamic visualization of 3D city models. Comput Environ Urban Syst 36(3):233–244

Open Geospatial Consortium (2012) OGC city geography markup language (CityGML) encoding standard 2.0.0

Open Geospatial Consortium (2012) OGC geography markup language (GML)—extended schemas and encoding rules 3.3.0

Salomon D (2007) Data compression—The complete reference, 4 edn. Springer London

Sony Computer Entertainment Inc. (2008): COLLADA—digital asset schema release 1.5.0. The Khronos group

Stadler A, Nagel C, König G, Kolbe TH (2009) Making interoperability persistent: a 3D geo database based on CityGML. In: Jiyeong L, Zlatanova S (eds) 3D geo-information sciences. Lecture Notes in Geoinformation and Cartography. Springer, Berlin, pp 175–192

Steinhage V, Behley J, Meisel S, Cremers AB (2010) Automated updating and maintenance of 3D city models. In: Peled A (ed) ISPRS joint workshop on "core spatial databases–updating, maintenance and services—from theory to practice", Haifa, Israel, pp 1–6

Stoter J, Visser T, van Oosterom P, Quak W, Bakker N (2011) A semantic-rich multi-scale information model for topography. Int J Geogr Inf Sci 25(5):739–763

XML Core Working Group (2010): XML Linking Language (XLink) Version 1.1 URL http://www.w3.org/TR/xlink11/

Zhang X, Ai T, Stoter J, Zhao X (2014) Data matching of building polygons at multiple map scales improved by contextual information and relaxation. ISPRS J Photogrammetry Remote Sens 92 (C) 98:147–163

Zhao J, Qingg Z, Du Z, Feng T, Zhang Y (2012) Mathematical morphology-based generalization of complex 3D building models incorporating semantic relationships. ISPRS J Photogrammetry Remote Sens 68(C) 68:95–111

Zheng Y, Cohen-Or D, Averkiou M, Mitra NJ (2014) Recurring part arrangements in shape collections. Compu Graph Forum 33(2):115–124

# Generalization of 3D IFC Building Models

**Andreas Geiger, Joachim Benner and Karl Heinz Haefele**

**Abstract** Today, Building Information Modeling (BIM) is mainly used in architecture. Typically, a BIM model contains detailed geometric and semantic information for design evaluation, simulation, and construction of the building. If, as on the regional and city levels, more than one building is considered, the information content of detailed BIM models might be too high. For applications like noise simulation or emergency management, representing buildings as block models, reduced outer-shell models or simplified indoor models are more suitable. Such models are typically found in Geospatial Information System (GIS) applications. This paper describes a process for BIM building models to extract different generalized representations for buildings and building elements. As an example, the definitions for such representations are based on the LoD concept of CityGML.

## 1 Introduction

Architectural building models are typically considered to be highly detailed. However, this is not the case at all stages of the design and construction process. In order to define and illustrate the characteristics of building models in different project phases, Levels of Development have been proposed by American Institute of Architects (AIA 2013).

These Levels of Development do not necessarily support the needs of building models in other areas of application and at other scales like, for example, in the case

A. Geiger (✉) · J. Benner · K.H. Haefele
Institute for Applied Computer Science, Karlsruhe Institute of Technology, Karlsruhe, Germany
e-mail: andreas.geiger@kit.edu

J. Benner
e-mail: joachim.benner@kit.edu

K.H. Haefele
e-mail: karl-heinz.haefele@kit.edu

© Springer International Publishing Switzerland 2015

M. Breunig et al. (eds.), *3D Geoinformation Science*,
Lecture Notes in Geoinformation and Cartography,
DOI 10.1007/978-3-319-12181-9_2

of geospatial data on the regional or city levels. Such applications often require domain-specific generalizations of 3D building data. One example of such a building model is the Building Module of the City Geography Markup Language (CityGML). It allows the representation of buildings at five dedicated Levels of Detail (LoD). With higher LoD, the geometric representation is refined and the semantic richness potentially increases.

CityGML buildings can be generated automatically, semi-automatically or manually, depending on the envisaged Level of Detail:

- LoD1—the building is geometrically represented as block model, where the footprint may be taken from cadastral maps and the height from laser scanning data or from cadastral data too.
- LoD2—based on a combination of 2D cadastral data and airborne laser scanning or photogrammetric data, the building's exterior shell is represented in a geometrically generalized way. Different parts of the exterior shell are semantically classified as e.g., wall or roof surfaces.
- LoD3—the exterior shell is represented in a geometrically detailed way. In addition to LoD2, doors and windows are identified in the exterior shell and represented as separate objects with surface geometry. This data may be extracted from terrestrial laser scanning data.
- LoD4—extension of the LoD3 model by an interior model. At the moment, this can only be done manually.

For existing buildings, automatic processes for generating LoD1 and LoD2 models are available. LoD3 and LoD4 models as well as any 3D model of a newly planned building have to be generated manually. In these cases, 3D BIM models are frequently available, but cannot be directly used for GIS applications. The central objective of this paper is to describe a generalization process transforming 3D BIM models into adequate GIS structures. The process is described on the basis of the CityGML standard, focusing on Levels of Detail 1–3, where only the building's exterior shell is represented. The transformation process starts using the Industry Foundation Classes (IFC) data format (Liebich 2007).

## 2 State of the Art of Generalizing 3D Building Models

The concept of generalization was originally introduced in 2D cartography. The International Cartographic Association (ICA) defines generalization as "the selection and simplified representation of detail appropriate to the scale and/or purpose of a map" (ICA 1973). Typical techniques are selection of the most important map features by simultaneously removing unnecessary details, simplifying or smoothing complex map features, and combination of small features. Many of these techniques are also relevant for the generalization of 3D building models.

For simplifying arbitrary (triangle) meshes, a number of general approaches exist. Many of them are based on special metrics for comparing different meshes

(Luebke et al. 2003). Other approaches for mesh simplification, being more suited to preserve the characteristic structures of buildings, are based on the extraction of geometrically defined features. During the simplification process, extracted features are evaluated and removed as necessary. Reviews of these approaches can be found in Babic et al. (2008) and Thakura et al. (2009).

Besides general mesh simplification, there also exist a number of specific approaches to generalizing 3D building models. Kada (2002) identifies characteristic building structures (e.g. coplanar, vertical or parallel parts) in a surface model, which are preserved in a merging process of surface parts. Thiemann and Sester (2004) segment a 3D building model, classify and process extracted segments and generate a generalized model by removing selected segments. In the approach by Kada (2006), a building model is transformed into a new representation based on half-spaces. The new representation supports a non-iterative process for eliminating "small" building parts. More approaches to generalizing 3D building models can be found in the review articles by Meng and Forberg (2006) and Sester (2007).

The techniques mentioned in the last paragraph are based on a purely geometric 3D building model. If data are available in a semantic data format like CityGML or IFC, additional information is available which can be used for the generalization process. Many such contributions use CityGML data of higher LoD (3 or 4) as starting point and try to transform them into LoD1 or LoD2 representations (Fan and Meng 2012; Baig and Rahmann 2013). Other authors discuss transformation and generalization processes based on IFC, which is also the central topic of this paper.

A first approach concerning a transformation from IFC building models into CityGML LoD1 was described by Nagel (2006) and Nagel and Häfele (2007). The CityGML concept according to LoD1 is characterized and strategies for the geometric transformation process are described. These methods are part of the generalization process presented here.

A solution for an IFC to CityGML LoD3 transformation in a three-stages generalization process is presented by Donkers (2013). Starting with a semantic filtering, relevant building elements are extracted. The second stage is the geometric transformation using Boolean and morphological operations like dilation and erosion to generate a volumetric representation of the complete building. The third stage called refinement is used to guarantee the compliance with ISO 19107 (ISO 19107:2003). The process is focused on LoD3 with an outlook to LoD4. LoD1 and LoD2 are not considered.

A mapping framework for transforming BIM to GIS in different LoD is presented by Cheng et al. (2013). To cover the semantic information, a new CityGML Application Domain Extension (ADE) called Semantic City Model was developed. The real geometric transformation seems not to be realized at the moment.

In Berlo (2011), the focus is set on the extension of CityGML with semantic IFC data. The GeoBIM-ADE uses IFC as source for GIS data.

An investigation is presented by El-Mekawy et al. (2012) concerning how much information of an IFC model can be transformed into a CityGML model. The disquisition is focused on the semantic information. It comes to the conclusion that a conversion implies data loss and thus a unified building model is proposed.

## 3 Building Models

The described generalization process is based on the open BIM standard IFC (Eastman 1999). If possible, all results are stored in the IFC data model as additional representations. In a second process, the results can be converted into the target format CityGML.

The Industry Foundation Classes (ISO 16739:2013) are an open standard for BIM, developed by buildingSMART (bSI 2014). It is based on STEP (ISO 10303) and the standardized data modelling language for product modelling EXPRESS (ISO 10303-11:2004), representing an entity-relationship model. Since version IFC4 it is an official international standard ISO 16739 (ISO 16739:2013). IFC defines two different encodings for the model data: The STEP Physical File (SPF) defined by ISO 10303-21 (ISO 10303-21:2002) and the STEP-XML defined by ISO 10303-28 (ISO 10303-28:2007). The most frequently used format is the SPF, having the advantage of compact size and being a human readable ASCII file format. The XML-based version is mostly used for exchanging partial models. Due to the large file size of the models, it is only used if interoperability with XML tools is required.

The version IFC2 × 3 contains 653 entities covering all phases of a building's life cycle. For Architecture, Engineering and Construction (AEC) modelling aspects, the default geometric representation of physical building elements is volumetric. This can be a parametric representation (e.g. extrusion of a parametric profile), a Constructive Solid Geometry (CSG) or a boundary representation (B-rep). IFC uses local Cartesian coordinates. For the site object, optionally a global geographic location can be specified. One or more buildings and building complexes can be represented with their complete building structure. The physical building elements are represented as objects with relations. Such relations can be used, e.g., for material information, properties or connections between building elements.

In IFC, a building element can have multiple geometric representations, which are differentiated by an identifier. There are pre-defined identifiers like, e.g., 'Body', 'Axis' or 'Footprint'. This concept is not comparable to the LoD concept of CityGML, because it mainly reflects modelling aspects of the corresponding AEC tools.

The City Geography Markup Language (CityGML) (Gröger et al. 2012) is an Open Geospatial Consortium (OGC) encoding standard for virtual 3D city models. CityGML is an application schema of the extensible Geography Markup Language (GML 3.1.1) (Cox et al. 2004). In contrast to 3D graphic formats, CityGML describes 3D object by their semantic meaning, their properties and their relations to other features. In order to cover major themes of a city, the standard is organized in 13 thematic modules (Gröger et al. 2012).

The most frequently used module of CityGML is the `Building module`, supporting five dedicated Levels of Details for representing buildings (see Sect. 1). A building can be structured into building parts. Exterior components like balconies or chimneys, which have a major impact on the outer characteristics of the building, are modelled as building installations. In LoD4, CityGML uses rooms as a spatial structure for the interior.

Buildings, building parts, rooms, and building installations can be represented by surfaces or solids. All these features can be semantically structured by, e.g., wall, roof or ground surfaces. Volumetric building elements like walls, roofs or slabs are not supported by CityGML.

IFC as well as the `Building` module of CityGML describe buildings as semantic objects with properties and relations and therefore can be considered as Building Information Models. Due to the intended application areas, used modelling techniques, and the history of the two organizations promoting the standards, significant differences between the models exist.

The most relevant difference between IFC and CityGML regarding data conversion is the modelling of buildings and building elements like walls, slabs or roofs. IFC usually uses `IfcBuilding` as pure container for building elements. Building elements are objects with properties, relations, and usually volumetric geometry representations. In contrast, CityGML allows `Building` to have an explicit solid or surface geometry. Building elements are not modeled as objects but as boundary surfaces for buildings or rooms without any further property or relations to other boundary surfaces.

This means that the volumetric representation of building elements in IFC has to be converted into boundary surfaces of the CityGML building or room. Figure 1 shows a simple example without considering the slabs of each storey and the different wall connections (butt joint, meter joint).

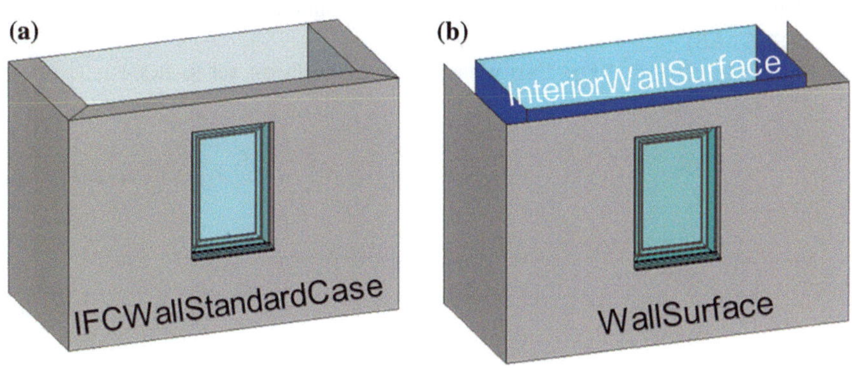


**Fig. 1** **a** IFC solid representation, and **b** CityGML surface representation

Another difference is the way building models are geo-referenced. While IFC is using local Cartesian coordination systems and with geographic coordinates on the site and the north direction, CityGML is using directly global coordinates regarding the given reference coordinate system. The conversion between the different geo-referencing methods can be done by well-known coordinate transformations.

# 4 Generalization Process

The goal of generalization is to reduce the geometric and semantic complexity of a building model without losing relevant information. As a first step, an intermediate data model, which is called `ExtrusionBaseModel`, is generated. Based on this model, LoD1 and LoD2 representations can be derived. LoD3 and (in the future) LoD4 are generated consecutively on the basis of the LoD2 model (see Fig. 2).

A suitable IFC model is prerequisite for a successful generalization. This means that the building model needs to have a closed exterior shell and is generally designed with walls, slabs, and roofs. The upper boundary of the building has to be modelled with slabs or roof elements. Additionally, horizontal slabs represent more or less the horizontal structure of the building with or without an explicitly defined storey structure. If the storey's structure is not defined in the data, it will be derived by interpreting slab elements.

## *4.1 Intermediate Data Model*

As a first step of the generalization process, an intermediate data model is generated (see Fig. 3) taking into account only relevant building elements (see Fig. 4). Currently, these building elements are:

- Walls (`IfcWall`, `IfcWallStandardCase`)
- Slabs (`IfcSlab` with `PredefinedType` undefined or set to `FLOOR` or `BASESLAB`)
- Roofs (`IfcRoof`, `IfcSlab` with `PredefinedType` set to `ROOF`)

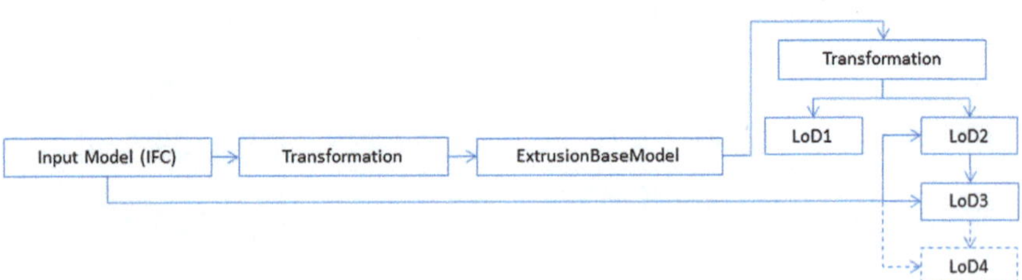


**Fig. 2** Schematic overview of the generalization process

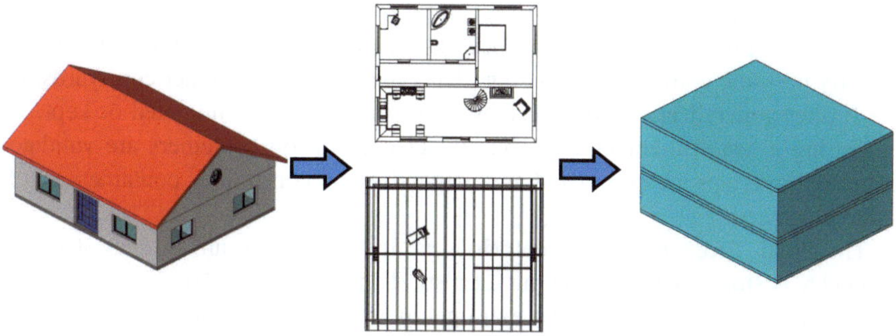

**Fig. 3** Transformation from IFC to the `ExtrusionBaseModel`

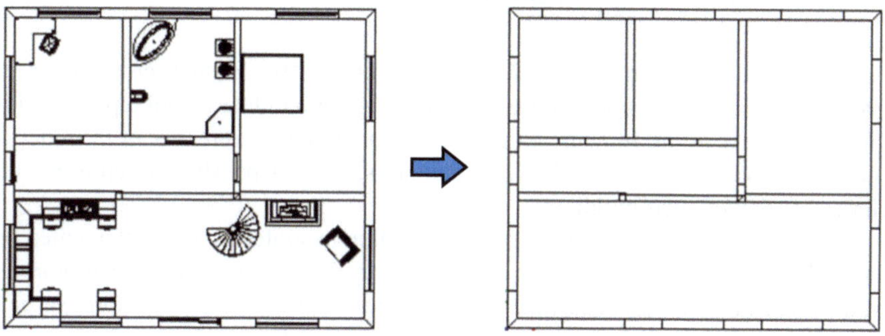

**Fig. 4** Semantic filtering of IFC data (*left* original model; *right* only IFCSLAB, IFCWALL, IFCBEAM)

- Facades (`IfcCurtainWall`)
- Beams and columns (`IfcBeam`, `IfcColumn`), optionally taken into account.

In IFC, different representation types are allowed and have to be considered for the transformation process. To get a uniform geometry base for the whole process, each considered building element is additionally represented by its footprint geometry. This is derived by a vertical projection of the original geometry into the x-y plane of the model coordinate system $(x, y, z) \rightarrow (x, y, 0)$.

Based on these footprints, extrusion containers are calculated. Each container aggregates information connected with a specific building element type taking into account the building structure, if possible. An extrusion container includes relations to the original IFC building elements, minimum and maximum z coordinates, the footprint geometry of the aggregated building elements, and further relevant information for post-processing the data.

Additionally, extrusion containers for building storeys are generated. If possible, they are derived from the building structure and combined with information from horizontal slab elements.

After completing the container generation, a collection of different extrusion containers is prepared for further processes. At first, heuristic methods are used to identify building components like, e.g., balconies, dormers or inner courtyards. For the further generalization process, these structures may be eliminated or kept.

During a last refinement step, the collected extrusion containers are validated. Identified overlaps are eliminated and, depending on constraints, containers can be split or merged.

The final `ExtrusionBaseModel` is the basis for all further steps. Its geometry is stored with the relating `IfcBuilding` and `IfcBuildingStorey` instances of the IFC model as an extra geometric representation.

## 4.2 Generation of the LoD1 Model

Following the definition of CityGML (Gröger et al. 2012; Benner et al. 2013a), a LoD1 model is a rough approximation of the original building model represented in one vertical extrusion. In many cases, this is not sufficient to represent the exterior shell of a building. Especially for landmarks, the shape should represent the characteristics of such buildings.

Based on the `ExtrusionBaseModel`, different strategies are implemented for the LoD1 generalization process. Depending on the strategy, building footprints, wall footprints or slab footprints are used to determine one or more vertical extrusions for the final LoD1 model (see also Sect. 5.2).

The first strategy calculates one extrusion for the whole building. Thereby, all footprints of the relevant building elements are used. The second variant focuses on walls. In this case, footprints of walls on same levels and same heights are combined and extruded. In the third case, footprints of horizontal slabs are used to create the extrusion geometries.

Because the geometric representation of the `ExtrusionBaseModel` is very close to the requirements of a LoD1 model (see Fig. 5), the main task of this transformation step is to adjust the vertical extrusions for the requested generation strategy. Optionally, it is also possible to create the resulting LoD1 geometries by an explicit storey structure. This means that for each storey, an extra extrusion is generated even if the footprints of the storeys are identical.

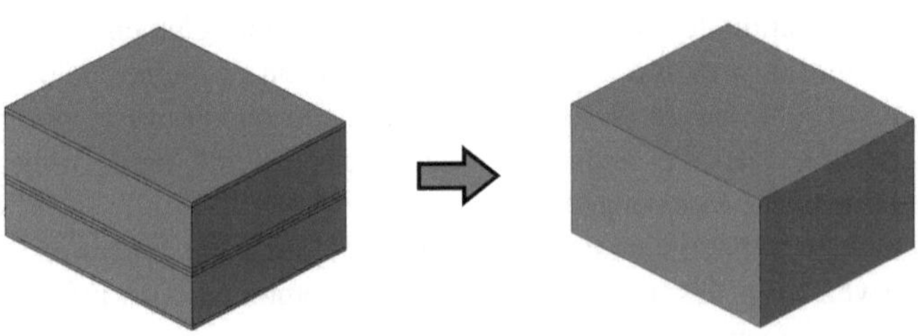

**Fig. 5** Transformation `ExtrusionBaseModel` into a LoD1 model

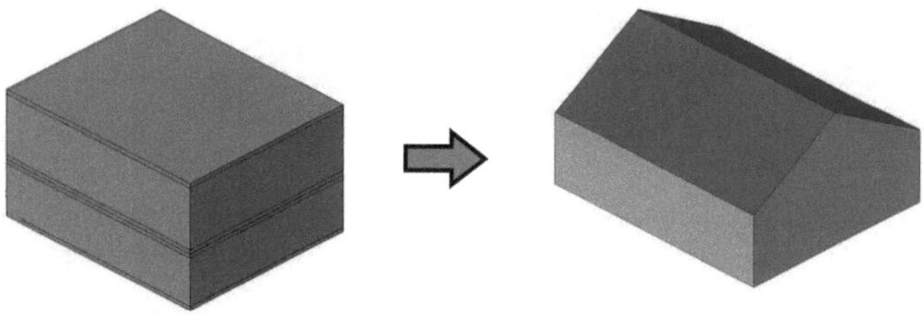

**Fig. 6** Transformation `ExtrusionBaseModel` into a LoD2 model

## *4.3 Generation of the LoD2 Model*

Compared to LoD1, the generation process for LoD2 is completely different. The basis is also the `ExtrusionBaseModel` but the generated outer contour of the building is much more detailed with a correct roof shape and a detailed classification of the outer boundary surfaces (see Fig. 6).

The first step is to assign the roof shape. An algorithm was developed to generate clipping planes for the extrusion containers of `ExtrusionBaseModel`, based on the original geometry of the IFC building elements `IfcRoof` and `IfcSlab` with `PredefinedType` set to `ROOF`. Supported geometry types in this process are extrusion and B-rep.

For B-rep geometries, all face normals are analyzed in order to identify upwards pointing faces. The outer polygons of these faces are transformed into bounded clipping planes.

For extrusion geometries, the procedure is different. Starting with the extrusion placement, the extrusion direction and its magnitude, the top plane of the extruded geometry can be calculated by a simple matrix operation. The boundary is derived from the extrusion profile, which is transformed into the plane.

In the second step, building elements are used for a semantic classification of the clipped extrusion geometries. For this, corresponding surfaces of IFC building elements and `ExtrusionBaseModel` objects are identified. In the classification mechanism, the following CityGML features are generated:

- `WallSurface`
- `OuterFloorSurface`
- `RoofSurface`
- `GroundSurface`
- `ClosureSurface`
- `BuildingInstallation`

While classifying the surfaces, an algorithm compares the distance of surfaces from IFC building elements to the generated surfaces of the `ExtrusionBase-Model`. The accuracy parameter `epsilon` for this distance test is adjustable, the default value it is set to is 50 mm. If the distance is smaller than `epsilon`, the corresponding surface is collected for further verifications. During this process,

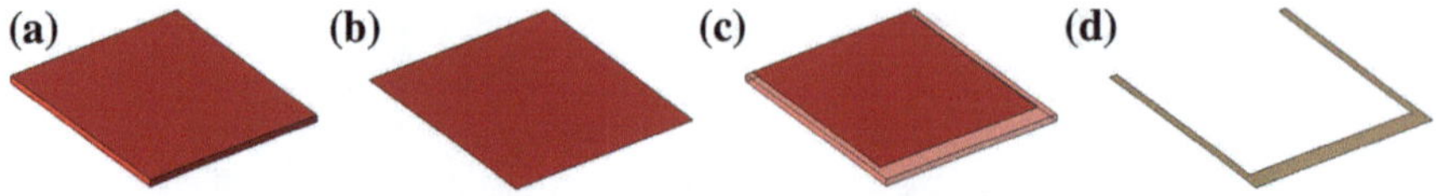


**Fig. 7** **a** IFC solid geometry, **b** CityGML roof surface, **c** subtraction of the roof surface with the top face of the IFC solid geometry, and **d** resulting geometry for roof overhangs

several faces of one building element, but also faces of different building elements can be found. All collected faces are compared in a second pass. In this pass, net area and the quantity of overlapping area, with corresponding surfaces of the `ExtrusionBaseMode` are checked to assign the correct type.

As an optional calculation step, the generation of roof overhangs can be initiated. The relevant geometry will be derived by a geometrical comparison of IFC roof elements with the created CityGML `RoofSurface`. By a projection of the CityGML `RoofSurface` on the IFC roof geometries, the overlapping parts are identified and separated by a geometric subtraction. For the CityGML model, these geometries are currently classified as `BuildingInstallation`. Figure 7 shows the steps for extracting roof overhangs.

## *4.4 Generation of the LoD3 Model*

The transformation process from a LoD2 model to a LoD3 model mainly differs in applying voids for doors, windows, and openings and creating the appropriate elements (see Fig. 8). In this step, the `ExtrusionBaseModel` is no longer used.

This process is realized by interpreting the relations of IFC. Cutouts are not explicitly given in the IFC geometry representations, but defined as relations between building elements, voiding elements, and a door or window elements. The first relation (`IfcRelVoidsElement`) describes a Boolean subtraction between, e.g., a wall and an opening element. Without handling this relation, the wall is given by their gross volume. The second relation (`IfcRelFillsElement`) is to place a door or window element within the hole of the subtracted opening element.

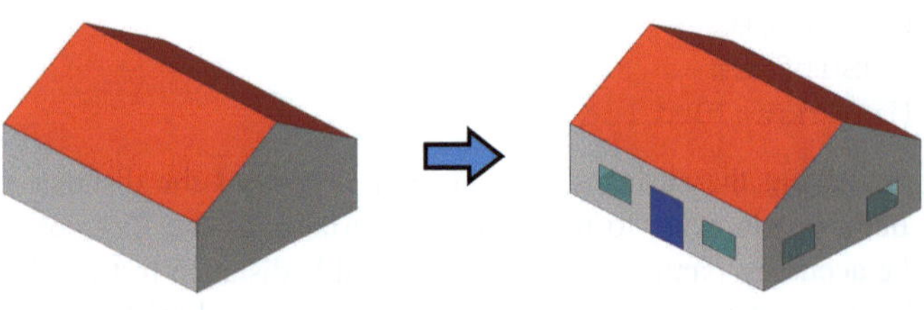

**Fig. 8** Transformation of a LoD2 into a LoD3 model

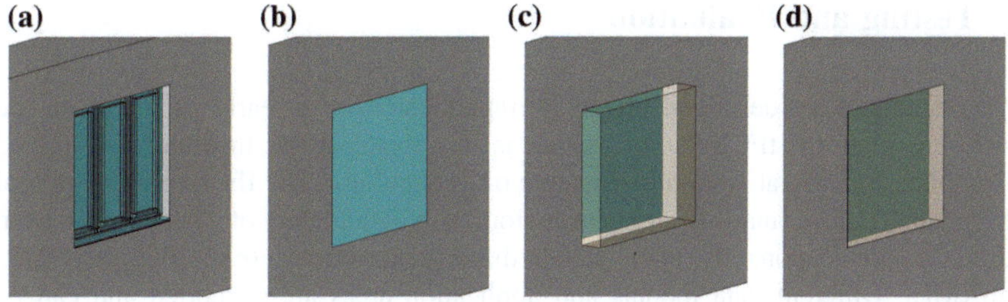


**Fig. 9** **a** IFC wall with window element, **b** CityGML `WallSurface` with window element in wall plane, **c** CityGML `WallSurface` with window element at correct position and IFC opening element (transparent), and **d** Embrasure realized by clipping opening element at wall and window plane

Doors and windows in IFC can be represented geometrically in different ways. Either by an explicit geometric representation, or by parameters describing operation type, lining, and panel. For non-rectangular elements, additionally a 2D profile should be given to describe the outer contour of the door or window. If only the parameters are given, the resulting geometry has to be calculated. Within the generalization process, the best solution to control the geometric complexity of these elements is to interpret the parameters, if available. Otherwise the geometry will be approximated by a projection of the opening element on the corresponding surface of the building element.

Unlike IFC, the opening elements in CityGML do not represent an additional element, but are already subtracted from the corresponding boundary surface. The door and window elements are defined as sub elements of the boundary surface and are geometrically located within the boundary surface. Openings without a door or window have to be represented as a `ClosureSurface` and cannot be related to any boundary surface.

For doors and windows, it is sufficient to reduce their geometry by a surface. This is different for embrasures. For representing a detailed façade, it is important to have the detailed depth and contour of the embrasures. Similar to roof overhangs, it is possible to activate the embrasure generation optionally. This step is currently under development. Basis is the `IfcOpeningElement` and the simplified door or window elements, created in the previous step. Clipping planes are derived from the door or window elements and from the boundary surface, e.g., `WallSurface`. Performing clipping operations with the geometry of the `IfcOpeningElement` results in a solid (negative mold of the embrasure), filling the recess between door or window surface and top edge of the boundary surface. Hence, the planes' normals are known, unneeded faces can be identified, and the remaining faces have to be inverted. The whole process is shown in Fig. 9. Finally, the surfaces of the embrasure have to be defined by the type of the boundary surface.

# 5 Testing and Evaluation

The methodology described above is implemented as an early prototype in the software platform IFCExplorer, which is developed at the Institute for Applied Computer Science at Karlsruhe Institute of Technology. The IFCExplorer is a tool for visualization, analysis, transformation, and integration of spatial data from different applications by open standardized data formats (e.g. IFC, cityGML, gbXML). Different data models and application areas are supported and can be opened either based on files or via standardized OGC web services (e.g. Web Feature Service, Web Map Service). For analyzing the different data models, a wide range of checking functionalities is available. The implemented functionality ranges from semantic validation, checking correctness of attributes, to geometrical and mathematical checks. This is the basis for realization of the here described generalization process (Benner et al. 2013b).

In order to prove the concept, single-family houses were used and transformed. The generated LoD models are internally represented as CityGML models and can be exported as CityGML 2.0 instance documents.

Each model is presented by an image of the IFC model and three images of the transformed LoD1, LoD2, and LoD3 model. Furthermore, for each model, a table shows all elements with a geometric representation, the types of the geometric representations, and the total amount of faces used for rendering.

## 5.1 Building Example 1

Example 1 is a simple free-standing building consisting of basement, ground floor, and roof floor. The LoD1 is represented in one vertical extrusion and for LoD2 and LoD3 the `WallSurface` is combined on all floors (Fig. 10; Table 1).

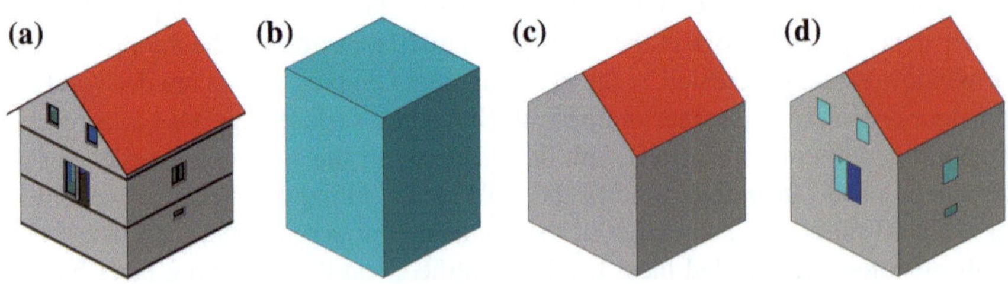


**Fig. 10** Example 1: **a** IFC Model, **b** CityGML LoD1, **c** CityGML LoD2, and **d** CityGML LoD3

**Table 1** Transformation results for building example 1

| | IFC | | | CityGML | LoD1 | LoD2 | LoD3 |
|---|---|---|---|---|---|---|---|
| Entities/Features | IfcDoor | 8 | → | Building | 1 | 2 | 2 |
| | IfcOpeningElement | 23 | | RoofSurface | | 4 | 4 |
| | IfcRailing | 1 | | WallSurface | | 1 | 1 |
| | IfcSlab[Floor] | 3 | | OuterFloorSurface | | | 2 |
| | IfcSlab[Roof] | 2 | | Door | | | 12 |
| | IfcSpace | 10 | | Windows | | | |
| | IfcStair | 2 | | | | | |
| | IfcWallStdCase | 19 | | | | | |
| | IfcWindow | 12 | | | | | |
| Geometry | B-rep | 261 | | MultiSurface | 1 | 7 | 21 |
| | Extrusion | 78 | | | | | |
| | SurfaceModel | 13 | | | | | |
| | Face | 4,849 | | Face | 6 | 7 | 21 |

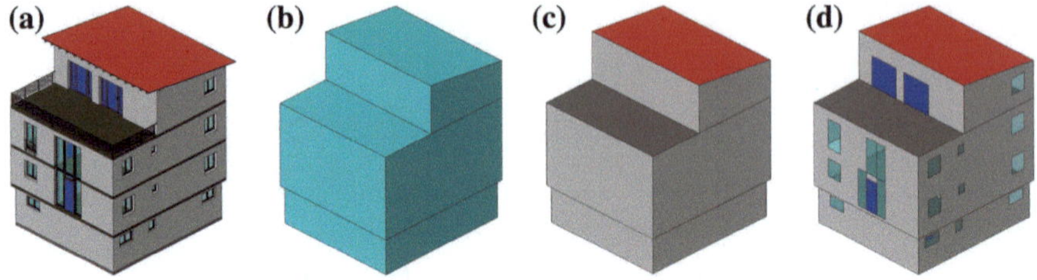


**Fig. 11** "Anwenderhandbuch": **a** IFC Model, **b** CityGML LoD1, **c** CityGML LoD2, **d** CityGML LoD3

### *5.2 Building Example 2*

The second building example is an official building model from the German "Anwenderhandbuch" which is a guide to the interoperability of CAD data from buildingSMART. For the LoD1 generation, the characteristics of the building are preserved by using the wall footprint strategy described in the LoD1 transformation step. The final geometry is represented in three vertical extrusions. LoD2 and LoD3 are generated with the same options, for this reason the `WallSurface` on ground floor and first floor are combined.

Due to an error in the internal Boolean operations, two of the window elements in the LoD3 model are missing in the front façade. The window elements are identified and created in the CityGML instance document but the geometric representation is missing (Fig. 11; Table 2).

## 6 Summary and Outlook

In order to use architectural building models like IFC in other application domains like geospatial environments or energy calculation, a method for semantical and geometrical generalization of IFC models is introduced. As a first example of this process, the transformation from IFC to CityGML (LoD1–3) is explained and first results are shown.

The method is implemented as an early prototype in the software IFCExplorer. The process is performed interactively and needs no further input. The results are immediately shown in the same document as the IFC model, which allows a direct comparison between the models. Tests have been carried out with simple single-family houses.

Next steps are implementing a prototype which can handle more complex buildings then single-family houses. This prototype will also include the consideration of overhanging parts of the building and the building's interior.

**Table 2** Transformation results for building example 2

| | IFC | | | CityGML | LoD1 | LoD2 | LoD3 |
|---|---|---|---|---|---|---|---|
| Entities/Features | IfcBeam | 17 | → | Building | 1 | 1 | 1 |
| | IfcDoor | 21 | | RoofSurface | | 12 | 12 |
| | IfcOpeningElement | 60 | | WallSurface | | 2 | 2 |
| | IfcRailing | 8 | | FloorSurface | | 1 | 1 |
| | IfcSlab[BaseSlab] | 1 | | GroundSurface | | | 4 |
| | IfcSlab[Floor] | 3 | | Door | | | 36 |
| | IfcSlab[Roof] | 2 | | Windows | | | |
| | IfcSpace | 21 | | | | | |
| | IfcStair | 6 | | | | | |
| | IfcWallStdCase | 38 | | | | | |
| | IfcWindow | 36 | | | | | |
| Geometry | B-rep | 1,089 | | MultiSurface | 1 | 16 | 56 |
| | Extrusion | 126 | | | | | |
| | Face | 17,426 | | Face | 16 | 16 | 56 |

With these experiences, the generalization process will be extended regarding other requirements for energy performance simulation. In this case, single-zone or multi-zones models will be created and exported to gbXML (2014).

## References

AIA (2013) Level of development specification version: 2013. Available at http://bimforum.org/wp-content/uploads/2013/08/2013-LOD-Specification.pdf. Accessed 2 July 2014

Babic B, Nesic N, Miljkovic Z (2008) A review of automated feature recognition with rule-based pattern recognition. Comput Ind 59(4):321–337

Baig SU, Rahmann AA (2013) A unified approach for 3D generalization of building models in CityGML. Int Arch Photogrammetry, Remote Sens Spat Inf Sci XL-4/W1:93–99

Benner J, Geiger A, Gröger G, Häfele K-H, Löwner M-O (2013a) Enhanced LoD concepts for virtual 3D city models. In: 8th 3D geoinfo conference, Istanbul

Benner J, Geiger A, Häfele K-H, Knüppel H (2013b) IFCExplorer—Ein Werkzeug für die Integration unterschiedlicher raumbezogener semantischer Daten. Geoinformatik, Heidelberg

bSI (2014) Available at http://www.buildingsmart.org/. Accessed 25 June 2014

van Berlo L (2011) Integration of BIM and GIS: the development of the CityGML GeoBIM extension. Advances in 3D Geo-Information Sciences. Springer, Berlin

Cheng J, Deng Y, Du Q (2013) Mapping between BIM models and 3D GIS city models of different levels of detail. In: 13th international conference on construction applications of virtual reality, London, 30–31 Oct 2013

Cox S, Daisey P, Lake R, Portele C, Whiteside A (2004) OpenGIS® geography markup language (GML) implementation specification, Version: 3.1.1, OGC 03-105r1. Open Geospatial Consortium

Donkers, S. (2013) Automatic generation of CityGML LoD3 building models from IFC models. MSc thesis, Delft University of Technology

El-Mekawy M, Östman A, Hijazi I (2012) An evaluation of IFC-CityGML unidirectional conversion. Int J Adv Comput Sci Appl 3(5):2012

Eastman CM (1999) Building product models: computer environments supporting design and construction. CRC Press, Boca Raton

Fan H, Meng L (2012) A three-step approach of simplifying 3D buildings modeled by CityGML. Int J Geogr Inf Sci 26(6):1091–1107

gbXML (2014) Available at http://www.gbxml.org/. Accessed 25 June 2014

Gröger G, Kolbe TH, Nagel C, Häfele K-H (2012) OGC city geography markup language (CityGML) encoding standard, Version 2.0, OGC doc no. 12-019. Open Geospatial Consortium

ICA—International Cartographic Association (1973) Multilingual dictionary of technical terms in cartography. Franz Steiner Verlag, Stuttgart

ISO 10303-11:2004 (2004) Industrial automation systems and integration—product data representation and exchange—part 11: description methods: The EXPRESS language reference manual

ISO 10303-21:2002 (2002) Industrial automation systems and integration—product data representation and exchange—part 21: implementation methods: clear text encoding of the exchange structure

ISO 10303-28:2007 (2007) Industrial automation systems and integration—product data representation and exchange—part 28: implementation methods: XML representations of EXPRESS schemas and data, using XML schemas

ISO 16739:2013 (2013) Industry foundation classes (IFC) for data sharing in the construction and facility management industries

ISO 19107:2003 (2003) Geographic information—spatial schema
Kada M (2002) Automatic generalization of 3D building models. Int Arch Photogrammetry, Remote Sens Spat Inf Sci (Part 4) 34: 243–248
Kada M (2006) 3D building generalization based on half-space modeling. In: Proceedings of the ISPRS workshop on multiple representation and interoperability of spatial data, Hannover
Liebich T (2007) Industry foundation classes—IFC2×3 documentation
Luebke D, Reddy M, Cohen JD, Varshney A, Watson B, Huebner R (2003) Level of detail for 3D graphics. Morgan Kaufmann, San Francisco
Meng L, Forberg A (2006) 3D building generalization. In: Mackaness W, Ruas A, Sarjakoski T (eds) Challenges in the portrayal of geographic information: issues of generalisation and multi scale representation. Elsevier Ltd, Amsterdam, pp 211–232
Nagel C (2006) Ableitung verschiedener Detailierungsstufen von IFC Gebäudemodellen. MSc thesis, University of Applied Science Karlsruhe, Research Center Karlsruhe
Nagel C, Häfele K-H (2007) Generierung von 3D-Stadtmodellen auf basis des IFC-Gebäudemodells. Entwicklerforum Geoinformatinstechnik 2007
Sester M (2007) 3D visualization and generalization. In: Proceedings of the 51st photogrammetric week, Stuttgart, 3–7 Sept 2007, pp 285–295
Thakura A, Banerjeea AG, Gupta SK (2009) A survey of CAD model simplification techniques for physics-based simulation applications. Comput Aided Des 41(2):65–80
Thiemann F, Sester M (2004) Segmentation of buildings for 3D-generalisation. In: Proceedings of the 7th ICA workshop on generalisation and multiple representation, Leicester, 20–21 Aug 2004 (on CD-ROM)

# Modeling and Managing Topology for 3-D Track Planning Applications

**Edgar Butwilowski, Andreas Thomsen, Martin Breunig, Paul V. Kuper and Mulhim Al-Doori**

**Abstract** Topology plays a central role in the modelling and management of n-d geo-information. However, hitherto the standardization of n-d topological data models is still at its beginning. A general model based on oriented hierarchical d-Generalised Maps is proposed to handle topology in object-oriented geo-database management systems. The implementation of the approach is demonstrated as module in DB4GeO, our service-oriented 3-d geo-database architecture. A 3-d tracks planning application example in the city of Munich is presented demonstrating the usage of the n-d topological model. Finally, an outlook on our ongoing 3-d geoinfo research in the Dubai region is given.

**Keywords** n-d topology · Geo-database · Tracks planning · 3-D planning application

## 1 Introduction

Navigation through parts of the geometry in 3-d city and infrastructure models is an important issue for analysing such models. Thus topological data models should play a central role in 3-d geo-information applications. However, to our knowledge,

E. Butwilowski (✉) · A. Thomsen · M. Breunig · P.V. Kuper
Karlsruhe Institute of Technology, Karlsruhe, Germany
e-mail: edgar.butwilowski@kit.edu

A. Thomsen
e-mail: andreas.thomsen@kit.edu

M. Breunig
e-mail: martin.breunig@kit.edu

P.V. Kuper
e-mail: kuper@kit.edu

M. Al-Doori
American University in Dubai, Dubai, United Arab Emirates
e-mail: maldoori@aud.edu

© Springer International Publishing Switzerland 2015

M. Breunig et al. (eds.), *3D Geoinformation Science*,
Lecture Notes in Geoinformation and Cartography,
DOI 10.1007/978-3-319-12181-9_3

the combined representation of topology in different spatial dimensions has not been investigated systematically.

In n-d space, a multitude of different representations from simplicial complexes to boundary representations and Constructive Solid Geometry are used (Schaeben et al. 2003; Van Oosterom and Stoter 2010). In this paper, we investigate how a more general concept, namely oriented hierarchical G-Maps, can be used to handle the topology of digital spatial models in a more generic way. The approach is general enough to support 2- and 3-dimensional models, as well as 2-d manifolds in 3-d space. An application example from a tracks planning project is presented.

## 2 Concept for Modeling N-D Topology

Cellular complexes and in particular cellular partitions of d-dimensional manifolds (d-CPM) are able to represent the topology of an extensive class of spatial objects (Mallet 2002). Based in algebraic topology, they provide a more general, less rigid framework than simplicial complexes. By stepwise aggregation of cells, hierarchies of d-CPM model a succession of generalisations (Köthe 2000). The topology of d-CPM can be represented by d-dimensional Cell-Tuple Structures (Brisson 1993) respectively d-dimensional Generalised Maps (d-G-Maps) (Lienhardt 1991). These possess the combinatorial structure of abstract simplicial complexes, each d-cell-tuple comprising d + 1 cells of different dimension, and related to its neighbours by involution operations. Lévy (1999) has shown that 3-d G-Maps have comparable space and time requirements as the well-known DCEL and radial edge structures, but a much wider range of application, and allow for a more concise and robust code. Lévy introduces hierarchical G-Maps (HG-Maps) for the representation of nested structures. Fradin et al. (2002, 2006) combine G-Maps with semantic information to model architectural complexes in a hierarchy of multi-partitions. G-Maps have been used to represent the topology of land-use changes (Raza and Kainz 1999) and are currently applied in the geoscientific 3-d modeling software GOCAD (Mallet 1992). Circular Incident Edge Lists (CIEL) is another useful data structure for the modelling of geoscientific subsurface data that has been introduced by Lévy et al. (2001).

## 3 Managing N-D Topology: Implementing Topology in DB4GEO

DB4GeO (Bär 2007; Breunig et al. 2010, 2012), our service-based geo-database architecture, has been designed to support geometric and topological applications in 3-d space. The core geometry model of DB4GeO is based on the model of simplicial complexes. In the model of simplicial complexes (for their application in GIS research see Egenhofer et al. 1989), geo-objects in 2-d and 3-d space are represented by a structure of connected non overlapping simplices, i.e. as triangles and tetrahedra, respectively.

In DB4GeO's implementation, an aggregation of contiguous non-overlapping simplex elements <Simplex>3DElt of the same dimension forms a simplicial complex represented by instances of the classes <Simplex>Net3DComp. The spatial part of a 3-d object may consist of several disjoint simplicial complexes. Therefore, the spatial part of a 3-d object in DB4GeO is represented by the classes <Simplex>Net3D, which may have multiple components. In order to provide efficient access to elements of a 3-d object, an R*-tree or an Octree is used for each net component. It is built up during the construction of the net component. As expected each element is enclosed by a minimum bounding box represented by the DB4GeO class MBB3D.

## *3.1 Boundary Representation and Level of Detail in DB4GeO*

Whereas the inherent structure of simplicial complexes serving to represent manifolds of dimension 0, …, 3 does not require a special treatment of topology, 3-d solids can also be represented as cellular complexes by their cell boundaries: Connected 3-d objects are decomposed into a number of connected cells, the faces, edges and nodes of which are in turn represented by simplicial complexes of lower dimension—triangle nets, polylines and point sets. The topological model of this BRep representation uses a variant of the HGMap structure proposed by Lévy (1999) (see below). In a way this cellular structure can be interpreted as a "generalization" of the non-manifold structure of its underlying components (Fig. 1).

In a similar way, simply connected groups of contiguous cells can be aggregated to a single cell by their common outer boundary, thus yielding a generalized representation of the solid. At each level, the topology of the cellular complex is described by a 3-GMap referencing geometrical boundary representation (Fig. 2). Fradin et al. (2006) provide a framework for the topology of such groups.

Consequently, in DB4GeO a hierarchy of G-Maps grouping comprehensive subsets of simplicial complexes is used to model the topologies of different levels of

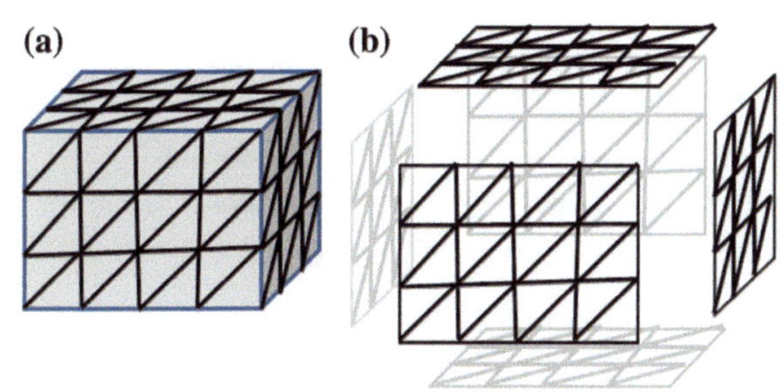


**Fig. 1** Boundary representation of a solid cell. **a** 3-D solid, and **b** 2D triangle net faces

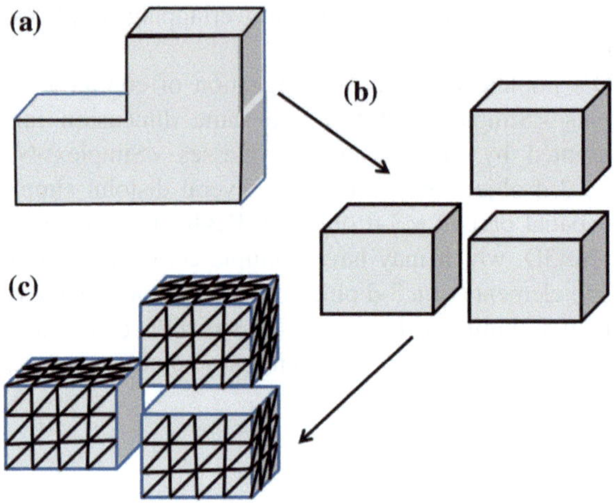


**Fig. 2** Aggregation of solid cells. **a** Aggregated cell, **b** Individual 3-d cells, and **c** 2D triangle net cell boundaries

detail in n-d models. DB4GeO supports spatial operations on G-Maps, e.g. the division of a cell by a shortest path that respects the meshing of the underlying simplicial complex.

The topology module of DB4GeO allows to freely navigate along the simple geometries of a complex geometry mesh. Therefore we developed a class model and directly connected it with the DB4GeO geometry kernel API. In our model, the class CellTuple plays a central role for the whole architecture: the CellTuple class is designed as a *composition* of cells for each dimension d, $0 \leq d \leq 3$ (Fig. 3).

A reference to exactly one cell of each dimension, i.e. to a Node, an Edge, a Face, and a Solid, is included. These cell classes are introduced into our model to represent cells that are *not restricted* by the constraints of the *simplicial* classes of DB4GeO: the cells of the G-Maps package need not to be simplices, they rather can be of any shape. Additionally, a CellTuple includes references *alpha*0, *alpha*1, *alpha*2, and *alpha*3 to four other CellTuple objects. These are involution transitions between cell-tuples that are explicitly modelled and stored in attributes of the class

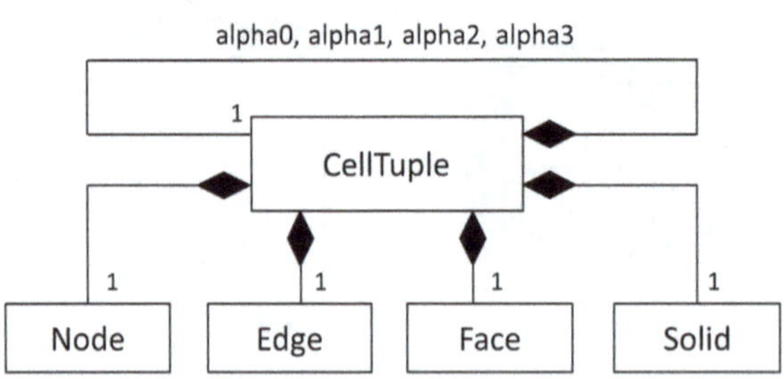


**Fig. 3** CellTuple class designed as composition of cell classes

field. Each involution $\alpha_i$ leads to a cell-tuple that is equal to the given cell-tuple with the difference that the cell of dimension *i* is exchanged.

In summary, our CellTuple class has the following definition in Java:

```
public class CellTuple {
    long id;
    Node node;     // reference to 0-cell
    Edge edge;     // reference to 1-cell
    Face face;     // reference to 2-cell
    Solid solid; // reference to 3-cell
    CellTuple alpha0; // alpha 0 involution
    CellTuple alpha1; // alpha 1 involution
    CellTuple alpha2; // alpha 2 involution
    CellTuple alpha3; // alpha 3 involution
}
```

All these references of a CellTuple object are set during a construction process. In the construction process, a cell net is build-up that consists of a set of cells and a set of cell-tuples. In our toolbox, the cells and cell-tuples are created on the basis of a simplicial complex.

The API user first creates a simplicial complex manually or loads data from a file or a network resource into DB4GeO. Then cells (i.e. Node, Edge, Face, and Solid objects) are created on the basis of the geometric and topological primitives of the simplicial complex. The primitive types that are provided by DB4GeO and that are used in this process are Point3D, Segment3D, Triangle3D, and Tetrahedron3D. All these primitives are geometrically embedded by x, y, z coordinates of Point3D class.

The geometric embedding (coordinates) of the simplicial complex is adopted by the cell complex. In a first step, the topological and geometric configuration of the simplicial complex is simply copied onto the cell complex, i.e. a Point3D object becomes a Node object, a Segment3D object becomes an Edge object, a Triangle3D object becomes a Face object and a Tetrahedron3D object becomes a Solid object. After the cell objects have been created, the CellTuple objects are build-up with all their references to the already existing cell objects. Then all missing information that is needed by the cell-tuple structure and that is not already directly available in the simplicial complex—like missing object references—are derived through interpretation of the simplicial complex.

In a second step, the cell complex is reduced and simplified in order to create a basis for further cell complex manipulation (this will be discussed in a section below).

As presented in Breunig et al. (2013a, b), the references between CellTuple objects and Cell objects are bi-directional (Fig. 4).

Each concrete cell class has a *getACellTupleOfCell()* method which returns an arbitrary cell-tuple of the cell (see *anyCellTuple* association in Fig. 4). A "cell-tuple of the cell" means a cell-tuple that vice versa contains the cell. With this model, it is easy to navigate from cell-tuple to cell and back at any point in the code. This is shown in the following example for the use case of iterating along the boundary of a face cell and printing all support points to standard out:

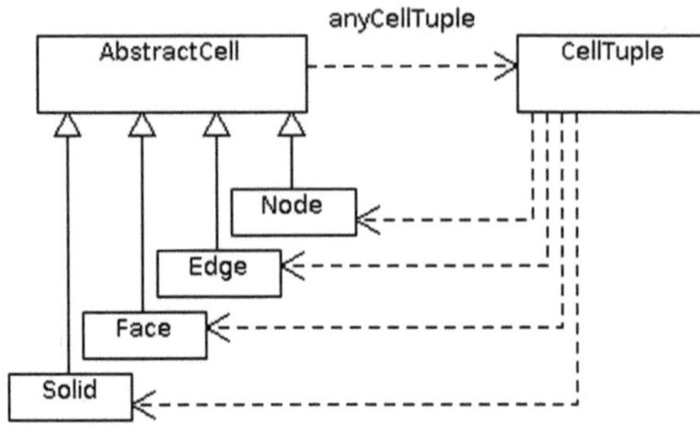


**Fig. 4** Bi-directional references between CellTuple and Cell. *Source* (Breunig et al. 2013a, b)

```
Face face = faceNetComponent.getFace(1);
CellTuple startCt =
        face.getACellTupleOfCell();
CellTuple currCt = startCt;
do {
  currCt = currCt.alpha0;
  System.out.println(currCt.node);
  currCt = currCt.alpha1;
} while (currCt != startCt);
```

In this coding session, the API user first queries a face net component for a particular face with the identifier (ID) 1. Proceeding with the returned face, the user receives an arbitrary cell-tuple of the face (*startCt*). The cell-tuple is used as a starting point to initiate a 2-orbit $(\langle \alpha_2 \rangle (startCt))$ "around" the face cell [an introduction to "orbits" can be found in Thomsen and Breunig (2007)]. In the do-while-loop, the $\alpha_0$ and $\alpha_1$ involutions are iterated rotatory each step. This is done by accessing the *alpha*0 and *alpha*1 field values of CellTuple class. A rotatory iteration of $\alpha_0$ and $\alpha_1$ involutions generates a 2-orbit, which navigates step-by-step along the nodes and edges of the face's boundary (Fig. 5).

The orbit stops and the program terminates when the cell-tuple of the current step (*currCt*) is the same as the start cell-tuple (which means the orbit is back to the beginning point).

Cells of the G-Maps package obey the requirements of Java Comparable interface (see Cell interface extending Comparable interface in Fig. 6).

This is helpful since it forces all realizing cell classes to implement a *compareTo ()* method. In our case, the method simply compares the IDs of two cells. Since the *compareTo()* method is realized, all cell objects can be stored in a Java *Set* or *Map* and retrieved efficiently by their ID, which is fast even at a high amount of cells.

The Cell interface specifies some methods that have to be realized by classes that claim to implement the interface. Some methods are specified in order to simplify the navigation on cell level. For example, the *getNeighbour…()* methods have to return all cells of dimension *j* that are neighboring a given cell of dimension *k*. For each dimension *j*, the Cell interface specifies an adapted method with a suitable cell type as the method's return type [e.g. *getNeighbourNodes()* or *getNeighbourEdges()*]. The methods have to return a list of all incident cells in the case that $j \neq k$ and a list of

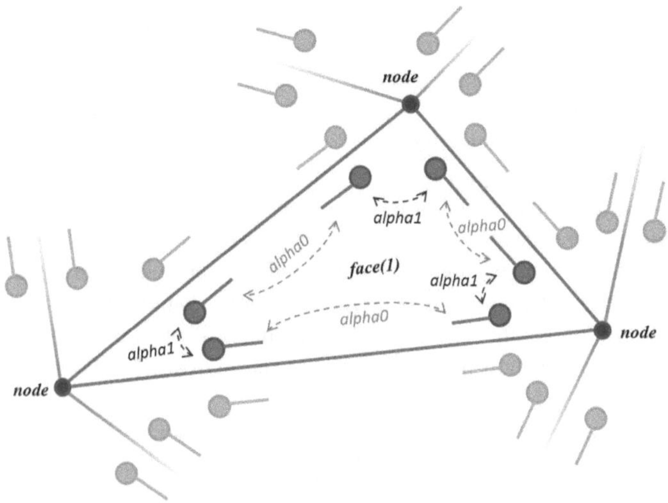


**Fig. 5** Example of a 2-orbit (*inside a triangular cell*). The cell-tuples are represented as *darts*: a *dart* positioned at *node(i)* along *edge(j)* within *face(k)* represents the cell-tuple *(node(i), edge(j), face(k))*. The set of all cell-tuples *{(node(), edge(), face(*1*))}* linked by sequences of involutions *alpha*0 and *alpha*1 describes the triangular *face(*1*)*

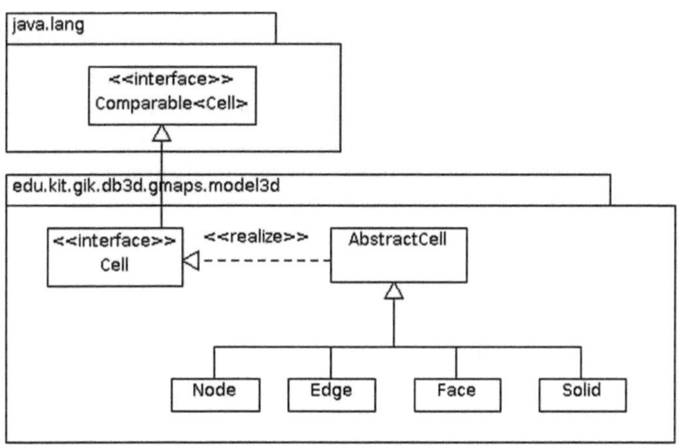


**Fig. 6** Class model of cells overview

adjacent cells in the case that $j = k$. For example, if invoking the *getNeighbourEdges ()* method on a Face object, then the method returns all edges that are incident to the face cell. In contrast, invoking the *getNeighbourEdges()* method on an Edge object, returns all edges that are adjacent to the given edge. As convenience methods, we also introduce the *isNeighbourOf(cell:Cell)* methods that have to take the cell that is given as the method's parameter value and to check whether the cell is adjacent or incident to the cell that the method is invoked on. Internally, the methods have to determine the cell's dimension and to invoke the appropriate *getNeighbour...()* method on the basis of the cell's dimension.

Similarly, we added *countNeighbour...()* convenience methods to the Cell interface that simply invoke the appropriate *getNeighbour...()* method and return the size of the result list. For example, the *countNeighbourNodes()* method of a Solid object returns the number of support points that the solid consists of, whereas the *countNeighbourSolids()* method of a Solid returns the number of its neighboring solids.

Additionally, the Cell interface demands a *getBoundary()* method that has to return the *d*-1-dimensional boundary cells of the given cell of dimension *d*. Internally, the *getBoundary()* method implementations simply have to invoke the appropriate *getNeighbour...()* method. For example, if invoked on an Edge object, the *getBoundary()* method internally invokes the *getNeighbourNodes()* method in order to receive *d*-1 boundary cells (nodes).

Of the presented methods of the Cell interface, only the *getBoundary()* method is individually realized for each concrete Cell class (Node, Edge, Face, and Solid). All other methods are implemented in an abstract cell class (AbstractCell), since their algorithms are equal for all different cell types. AbstractCell class is the superclass of all concrete cell classes.

The objective for the development of the GMaps package is to provide both advanced (fast and flexible) navigation capabilities inside the already available net structure of the simplicial complex data types of DB4GeO, as well as the possibility to define new cell types that are more general than simplices.

In order to integrate the already available functionality of the DB4GeO kernel with the model of general cell types, we implement the GMaps package as a *two level hierarchy*, where the general cell types are defined as the *object level* and the simplicial complex data types as the underlying *net level* (Fig. 7).

We then extend the CellTuple definition by two field attributes *higher* and *lower* of type CellTuple. These form the type-reflexive, bi-directional references between cell-tuples of lower level of detail (object level) and higher level of detail (net level).

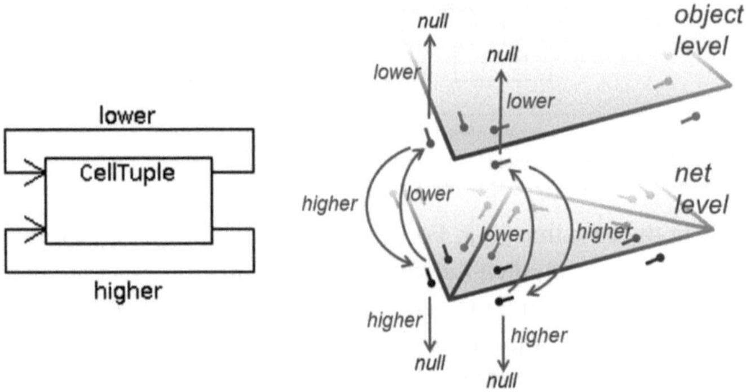


**Fig. 7** Cell-tuples of object level and net level are bi-directionally linked by higher and lower references

In DB4GeO, the construction process for G-Maps structure performs in two steps: In a first step, all cell-tuples on net level are created on the basis of the underlying simplicial complex. Second, the object level is derived from the net level. In this step, only the cell-tuples of the components' boundaries are considered: for each cell-tuple at net level boundary, a sibling is created at object level. Each original net level cell-tuple (*ctNL*) is linked to its newly created sibling at object level (*ctOL*), in a way that *ctNL.lower* = *ctOL* and back-referenced (*ctOL.higher* = *ctNL*). If we already are on object level, then the return value of *lower* attribute is *null*. Vice versa, if we already are on net level, then the return value of *higher* attribute is also *null*. By this, we can always check whether we are on object level or on net level by checking *ct. lower* == *null* (object level) and *ct.higher* == *null* (net level). Every cell-tuple at object level has a valid reference to its representation at net level (i.e. *ctOL. higher* == *null* is never the case). Instead, cell-tuples at net level may or may not have a representation at object level, depending on their embedding. If they have no object level representation, i.e. if a certain detail is missing at the object level, then *ctNL. lower* == *null* is true.

The presented concept gives us free navigation capabilities between net level and object level and is flexible enough to be extended by any number of additional levels of detail.

The boundary representation (object level) of a geo-object can be used as a starting point to create additional cells at object level through suitable topological editing operations. The G-Maps package user can insert or delete nodes, edges, and faces. For example, by inserting an edge into a face, the face cell is split into two new face cells. In such case, the G-Maps API takes care of creating all new cell and cell-tuples and creating and maintaining all bi-directional multi-level references. The internal method for edge creation depends on the application purpose and can be exchanged. As a prototype, we implemented an edge creation mechanism on the basis of the shortest path between two nodes of the net level mesh, following triangle boundaries.

With the presented model, it is easy to navigate not only on top of meshed data like simplicial complexes but also to freely navigate between the object level and net level representation of a geo-object. In practice, this can be used for efficient algorithms in hybrid models that integrate e.g. TIN and B-Rep in one application.

# 4 Tracks Planning Example

## *4.1 Handling Levels of Detail*

The planning of a system of tunnels for underground railways is a complex task with different protagonists such as city planners, civil engineers, geologists etc. However, they share geometry and topology as central parts of their models. Our application example, an ongoing 3-d tracks planning project in the city of Munich, investigated by the research group "Computer-supported Cooperative Planning for

| | Level 1 | Level 2 | Level 3 | Level 4 | Level 5 |
|---|---|---|---|---|---|
| **Subway Planning / Tunnel section** | 2D polyline | 2D polyline (with width) | 3D Tunnel geometry | precise 3D geometry (with inner shell, drainage, separating wall) | precise 3D geometry incl. rails /, traffic lanes, fan equipment, cables, lighting devices, traffic signs |
| **Figure** | 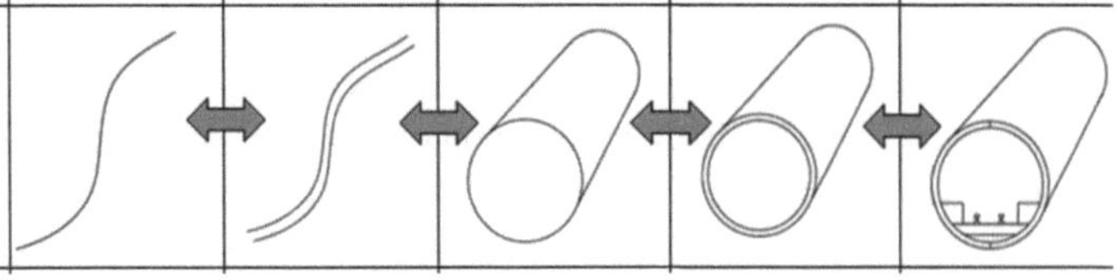 | | | | |

**Fig. 8** Geometric objects on different LoD in tunnel planning (by André Borrmann, TU Munich)

3-d City and Building Models" (*3DTracks*) (Breunig et al. 2011), demonstrates how topology is used to navigate through 3-d planning models: In subway planning, completely different scales have to be considered—ranging from the scale of several kilometers for the general design of the subway alignment down to centimeter scale for the detailed planning of traffic nodes.

The use of multi-scale models is well established in the Geographical Information System (GIS) domain (van Oosterom and Schenkelaars 1995), and forms an integral part of the respective data exchange standards, such as CityGML for example Kolbe (2008). However, these approaches focus on static models and primarily aim at supporting the visualization of the models. Accordingly, such approaches are less suitable for multi-scale models used in planning processes where frequent modifications result in high dynamics. Currently there is yet no well-established methodology available to ensure and preserve consistency between such dynamic multi-scale models. In a first step, the number of levels-of-detail (LoD) as well as the corresponding geometrical representation on each LoD has to be defined (Fig. 8).

Contrary to the bottom-up definition of levels-of-detail by successive generalizations in GIS, in building planning LoDs are established top-down starting with a general outline and stepwise adding detail. On LoD1, the tunnel is represented by only its main axis (the *track line*), while on LoD5, it is represented by a precise 3-d model including all planning details. The representations on the diverse LoDs result from different detailing demands in the individual planning stages. In a second step, half-automatic transformations between the geometric representations at different LoDs should be provided.

In general, the storage and management of multiple representations of real world entities involve logical redundancies which have to be carefully handled in order to avoid inconsistencies. This applies in particular when model modifications are as frequent as in an ongoing planning process. In the context of the *3DTracks* project, a logical framework has been developed for the multi-representation of tunnel planning models at several LoDs. Two sub-projects of the *3Dtracks* research group at Munich University of Technology have developed a methodology and software system that aims at ensuring consistency across levels-of-detail during synchronous co-operative interactive tunnel planning (Borrmann et al. 2013; Borrmann and Jubierre 2013). A characteristic feature of their level-of-detail hierarchy is the interdependency of semantics, spatial properties and consistency constraints. Hence they employ the combination of a semantic model *SM* with an associated procedural (CSG) geometry model *PM*. The *SM* describes the functions of objects and the dependencies between different model parts, in particular the relationships between LoDs whereas in the *PM*, the progression from lower towards higher LoDs is realized by successively applying geometric operations adding more and more detail.

Borrmann and Jubierre (2013) define 5 levels-of-detail of a tunnel segment (Fig. 8): LoD1: track line, LoD2: tunnel volume as a compact solid, LoD3: tunnel wall, LoD4: tunnel wall with track bed, LoD5: Tunnel with full equipment. A number of constraints linking the objects ensure spatial consistency: the inner "hollow" volume of the tunnel must be completely contained within the volume defined by the outer hull, contiguous tunnel parts must exactly fit together etc. Because of such constraints and other dependency relations, the directed graph model of LoDs consists of a stratified tree augmented at each LoD by links between the nodes of different branches.

In the hybrid model of tunnel planning, the LoD hierarchy is established within the semantic model *SM* and carries over to the spatial procedural model *PM* by the application of a combination of spatial operations and spatial constraints. A closer look shows that semantic, topological and geometrical aspects cannot be strictly separated, even though each aspect follows its own rules: e.g. the semantics of a tunnel require, that contiguous tunnel parts fit together at all LoDs, hence their topological models must be connected, and the geometries of their front parts must match. Parts of the tunnel equipment at LoD5 like e.g. railway signals are topologically disconnected separate units the geometrical position of with however is determined by engineer standards.

For application purposes—like augmented reality visualization (cf. Urban et al. 2013) and integration with city models (cf. Steuer et al. 2013)—from the *PM* a corresponding explicit representation *EM* at different LoDs is derived by means of CAD software (Fig. 9). The transformation into an explicit representation *EM* of the model geometry does only rudimentarily preserve these logical dependencies. Instead we must try to reconstruct them by means of the references between semantic and procedural model parts and their explicit representation. From a DBMS point of view, these structures and dependencies are to be taken for granted and must be represented as such in the hybrid database.

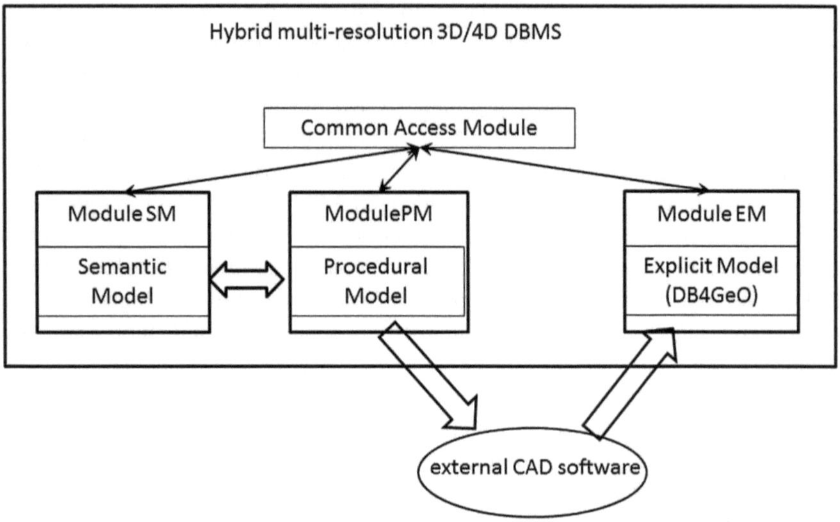


**Fig. 9** Modifications must take place on the semantic *SM* and procedural *PM* representations. External CAD software (AutoDesk Inventor) is used to transform the updated version into a new explicit model. There is no feedback of modifications from *EM* to *PM* and *SM*

## *4.2 A Hybrid Multi-representation 3D/4D DBMS for the 3DTracks Tunnel Planning Project*

This leads to the concept of a hybrid, multi-representation 3D/4D database for the management of different temporal versions of 3-d geometry models with associated semantics and derived explicit representations at different levels of detail (Fig. 10). Its data storage comprises four different components—*SM* for model semantics and dependencies, *PM* for its procedural (CSG) geometry, *EM* for an explicit representation, *AM* for additional information (annotation, comments etc.). A common *access module* supports navigation across the whole database, and delegates elementary queries stemming from the query module, to the corresponding storage modules.

From the way the data are generated, it follows that there are strong dependencies between the different representations. In particular, the *EM* is completely dependent on *PM* and *SM,* and there is no feed-back from *EM* to *PM*. In consequence, all updates of the data set must first take place on *SM* and *PM*, and then be carried over to *EM* (Fig. 9). For the implementation of the *EM* component, an adapted version of DB4GeO is to be used. At each level of detail, the topology of the explicit representation of the model parts on the object level consists of cellular complexes in BRep referencing the 2-d simplicial complexes on the net level (Fig. 11).

For the navigation on the hybrid database, the access module uses a graph of the dependencies derived from the *SM* and *PM*, and enhanced by spatial information from the *EM* like e.g. bounding boxes and by thematic information from the *AM* like material, costs etc.

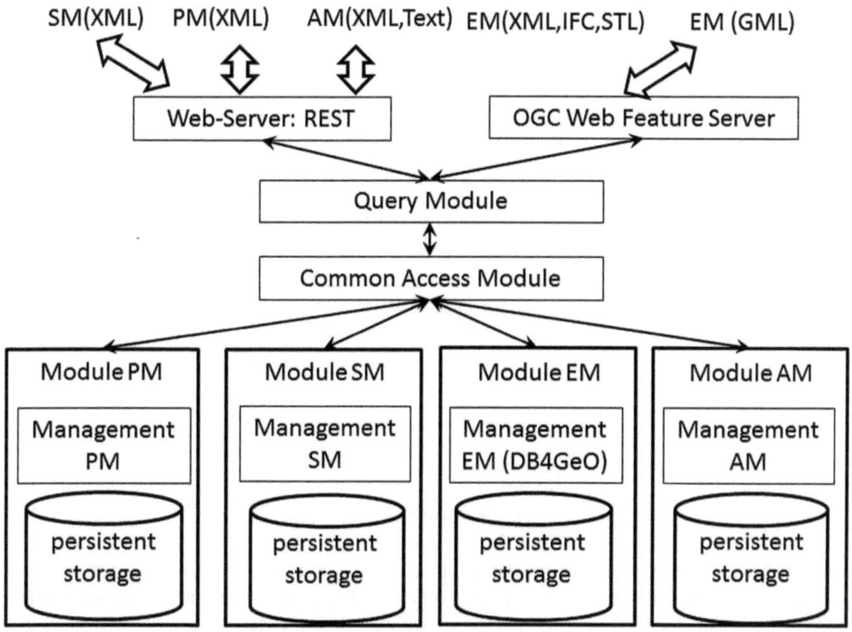


**Fig. 10** Architecture of a hybrid spatio-temporal database. Each model representation is stored in a separate database module (*PM* procedural, *SM* semantic, *EM* explicit, *AM* annotation). Integrated access is provided by a common access module

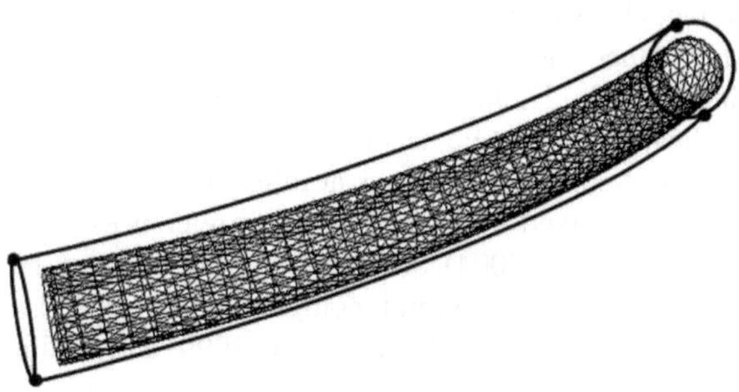

**Fig. 11** Explicit representation of the topology of the outer hull of a tunnel part at LoD1. For the topology at the object level, a cellular complex of only two faces is sufficient referencing two triangle nets at net level

Whereas the hierarchy of the Levels of Detail is defined within the semantic model *SM* by references and cross-references between objects of *SM* and the objects of *PM* and *EM*, its logical structure carries over to these representations (Fig. 12).

To support the navigation across different LoDs, there are essentially two ways of handling dependencies between different representations of a spatial object.

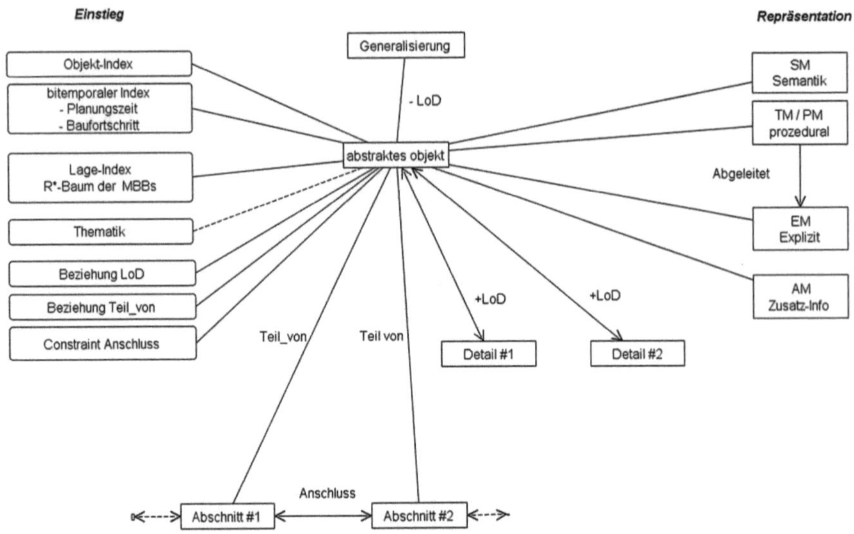


**Fig. 12** Multi-representation of a spatial object in the hybrid database. An abstract object references semantic (*SM*), procedural (*PM*), explicit (*EM*) representation and annotation (*AM*), at different levels of detail. Relationships and constraints describe dependencies between objects. Access is provided by different spatiotemporal and thematic indices

On the one hand, a hierarchy of LoDs can be established ad hoc by applying a combination of well defined generalization resp. specialization operations to an explicit spatial model. An important question is whether such operations are reversible, and whether they are Eulerian—i.e. preserve essential topological properties (Mäntylä 1988; Thomsen and Breunig 2007). A particular example of reversible generalization of triangle nets is the *progressive meshes* method (Hoppe 1996). On the other hand, representations at different LoDs can be produced separately, and afterwards linked together by explicit references (Haunert and Sester 2005; Anders and Bobrich 2004). The two approaches can be combined. In DB4GeO, a close relationship between topologies at different LoDs is provided by the explicit linking of cell-tuples, following Lévy (1999) and Fradin et al. (2002). If we try to apply this approach to the *EM* component of the hybrid DBMS, however, the question arises whether the two LoD hierarchies of *SM/PM* and *EM* can be made compatible (Fig. 13). If this were the case, navigation across LoDs on the explicit part of the DBMS might be enhanced by topological and geometrical operations, instead at each step passing by cross-references to the semantic model and back again. However, as the two hierarchies are established independently, it seems to be impossible to achieve compatibility, because of the unidirectional dependency of *EM* on *SM/PM*. The only generalization operation that can be performed independently on the EM is a change of the mesh densities of the boundaries elements, as this parameter is not controlled by the semantic and the procedural model. Mesh density, however, does not appear in the definition of

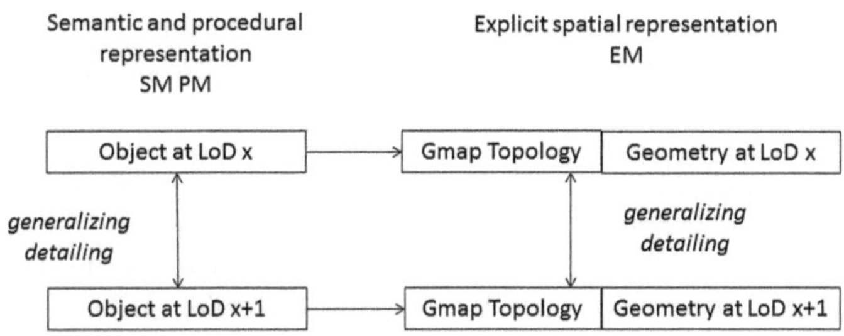


**Fig. 13** Combination of semantic and spatial LoD hierarchies

LoDs mentioned above. Therefore, for the navigation on LoDs, the access module uses the dependencies from the *SM* and *PM*, combined with references to the objects of the *EM*. It has yet to be investigated if any gain in efficiency can be obtained by establishing links between LoDs within *EM* that are completely controlled by the dependencies within the semantic and procedural models.

# 5 Conclusion and Outlook

In this paper we have investigated an approach for the modelling and management of n-d topology in geo-database management systems. As an implementation example we demonstrated topology classes and their relationships to geometry classes of DB4GeO, our service-oriented geo-database architecture. Finally we applied the introduced topology approach to topological objects of a 3-d tracks planning example in the city of Munich, Germany.

In our future research work we plan to use new application examples in the Dubai region that is generating a strong requirement for geospatial services for above- and sub-surface models for applications such as 3-d city modelling, 3-d tracks planning, oil exploration, etc.

Problems to be solved in the region are water sources seriously threatened by pollution, agriculture intensification, decline of urban settlements and transport networks, soil sealing and fragmentation of landscape. A 3-d model covering a vast area of the Emirates, including the city itself and neighbouring territories to north, west and south, will be able to create new kinds of database queries helping to understand and analyse urban and sub-surface planning.

Apart from conventional 3-d applications there is also an increasing demand of mobile systems. For instance in May 2013 the vice-president and prime minister of the UAE and ruler of Dubai, Sheikh Mohammed bin Rashid Al Maktoum, announced the "Mobile Government" initiative supporting the development of mobile systems. In this context mobile GIS are of prime importance e.g. the GIS department at Dubai Municipality is currently working on mobile applications like Dubai Cadastral, Dubai Map und Dubai Utilities Map.

Until now most of the mobile GIS applications are handling 2-d data following the Simple Feature guidelines. But the applications capable of handling 3-d geodata e.g. sub-surface or city models are on the rise. The area of augmented reality, especially in the context of Smart Cities has enormous potential.

At the Karlsruhe Institute of Technology (KIT) an Android based augmented reality viewer was developed in 2013 which is capable of presenting 3-d building models (Breunig et al. 2013a, b). For the further development in this research area still many challenges need to be solved. Precise location management, limited processing power and limited memory of the mobile devices are just some of the problems which must be resolved.

The long-term goal is to contribute to the enhanced planning, designing and development processes being undertaken as part of the Dubai Vision for the next two decades.

**Acknowledgments** This work is funded by the German Research Foundation (DFG) within the Research group 1546 "Computer-Aided Collaborative Subway Track Planning in Multi-Scale 3D City and Building Models", grant no. BR 2128/14-2.

## References

Anders K-H, Bobrich J(2004) MRDB approach for automatic incremental update. In: ICA workshop on generalisation and multiple representation, Leicester

Bär W (2007) Management of geoscientific 3D data in mobile database systems. In: German. Ph. D. thesis, University of Osnabrück, Germany, 166 p

Borrmann A, Jubierre JR (2013) A multi-scale tunnel product model providing coherent geometry and semantics. In: Proceedings of the 2013 ASCE international workshop on computing in civil engineering, Los Angeles

Borrmann A, Flurl M, Jubierre J, Mundani R-P, Rank E (2013, submitted) Synchronous collaborative tunnel design based on multi-scale infrastructure models. Submitted to advanced engineering informatics. Elsevier,Amsterdam

Breunig M, Schilberg B, Thomsen A, Kuper PV, Jahn M, Butwilowski E (2010) DB4GeO, a 3D/4D geodatabase and its application for the analysis of landslides. In: Geographic information and cartography for risk and crisis management. Lecture notes in geoinformation and cartography. Springer, Berlin, pp 83–102

Breunig M, Hinz S, Menninghaus M, Vögtle Th, Rank E, Schilcher M, Borrmann A, Mundani R-P, Ji Y, Donaubauer A, Steuer H (2011) Towards computer-aided collaborative subway track planning in multi-scale 3d city and building models. In: Proceedings 3D geoInfo 2011, Wuhan, 9 p

Breunig M, Butwilowski E, Golovko D, Kuper PV, Menninghaus M, Thomsen A (2012) Advancing DB4GeO. In: Proceedings 3D Geoinfo 2012, Quebec, Canada. Lecture notes in geoinformation and cartography. Springer, pp 193–210

Breunig M, Butwilowski E, Kuper PV, Golovko D, Thomsen A (2013a) Topological and geometric data handling for time-dependent geo-objects realized in DB4GEO. In: Timpf S, Laube P (eds) Advances in spatial data handling. Advances in geographic information science. Springer, Berlin, pp 1–13

Breunig M, Kuper PV, Dittrich A, Wild P, Butwilowski E, Al-Doori M (2013b) Design, implementation and applications of 3d web-services in DB4GEO. ISPRS Ann photogrammetry Remote Sens Spat Inf sci II-2/W1:87–94. doi:10.5194/isprsannals-II-2-W1-87-2013

Brisson E (1993) Representing geometric structures in d dimensions: topology and order. Discrete Comput Geom 9:387–426

Egenhofer MJ, Frank AU, Jackson JP (1989) A topological data model for spatial databases. In: Proceedings ACM SIGMOD, LNCS 409. Springer, Berlin

Fradin D, Meneveaux D, Lienhardt P (2002) Partition de l'espace et hiérarchie de cartes généralisées. In: AFIG 2002, Lyon, décembre 2002, 12 p

Fradin D, Meneveaux D, Lienhardt P (2006) A hierarchical topology-based model for handling complex indoor scenes. Comput Graph Forum 25(2):149–162

Haunert J-H, Sester M (2005) Propagating updates between linked datasets of different scales. In: Proceedings of XXII international cartographic conference, A Coruña, Spain, 11–16 July

Hoppe H (1996) Progressive meshes. In: ACM SIGGRAPH 1996, pp 99–108

Kolbe TH (2008) Representing and exchanging 3D city models with cityGML. In: Lee J, Zlatanova S (eds) Proceedings of the 3rd international workshop on 3D geo-information, Seoul, Korea. Lecture notes in geoinformation and cartography, Chap. 2, Springer, Berlin

Köthe U (2000) Generische Programmierung für die Bildverarbeitung. Dissertation, FB Informatik, Universität Hamburg, 274 p

Lévy B (1999) Topologie Algorithmique: Combinatoire et Plongement. Ph.D. thesis, INPL Nancy, 202 p

Lévy B, Caumon G, Conreaux S, Cavin X (2001) Circular incident edge lists: a data structure for rendering complex unstructured grids. In: Proceedings of visualization 2001, (VIS'01). IEEE, San Diego, pp 191–557

Lienhardt P (1991) Topological models for boundary representation: a comparison with n-dimensional generalized maps. Comput Aided Des 23(1):59–82

Mallet JL (1992) GOCAD: a computer aided design programme for geological applications. In: Turner AK (ed) Three-dimensional modelling with geoscientific information systems, NATO ASI 354. Kluwer Academic Publishers, Dordrecht, pp 123–142

Mallet JL (2002) Geomodelling. Oxford University Press, Oxford, 599 p

Mäntylä, M. 1988. An Introduction to Solid Modeling. Computer Science Press, 401 p

Raza A, Kainz W (1999) An object-oriented approach for modelling urban land-use changes. ACM-GIS 1999:20–25

Schaeben H, Apel M, vd Boogart G, Kroner U (2003) GIS 2D, 3D, 4D, nD. Informatik Spektrum 26(3):173–179

Steuer H, Donaubauer A, Kolbe TH, Flurl M, Mundani R-P, Rank E (2013) Planning inner-city-railway-tracks: dynamic integration of geospatial web services in a collaborative multi-scale geometry modelling environment. In: Proceedings of 20th international workshop: intelligent computing in engineering, Vienna, 1–3 July

Thomsen A, Breunig M (2007) Some Remarks on Topological Abstraction in Multi Representation Databases. In: Popovich VV, Schrenk M, Korolenko KV (eds) Proceedings of 3rd international workshop on information fusion and geographical information systems IF&GIS-07, St. Petersburg. Lecture notes in geoinformation and cartography. Springer, Berlin, pp 234–51

Urban S, Leitloff J, Wursthorn S, Hinz S (2013) Self-localization of a multi-fisheye camera based augmented reality system in textureless 3D building models. ISPRS Ann Photogramm Remote Sens Spatial Inf Sci II-3/W2:43–48

Van Oosterom P, Schenkelaars V (1995) The development of an interactive multi-scale GIS. Int J Geogr Inf Sys 9(5):489–507

Van Oosterom P, Stoter JE (2010) 5D data modelling: full integration of 2D/3D space, time and scale dimensions. In: GIScience 2010, pp 310–324

# Multi-resolution Models: Recent Progress in Coupling 3D Geometry to Environmental Numerical Simulation

**Vasco Varduhn, Ralf-Peter Mundani and Ernst Rank**

**Abstract** In this paper, we present recent results on coupling geographic, infrastructure, and building models to multi-resolution numerical simulations. In order to achieve this, a parallel data access framework with interfaces to all parts of the simulation pipeline such as pre-processing, numerical simulation, and post-processing has been developed. The applicability of the approach presented in this work is shown by simulating urban flooding including surface flow of a city, the pipe network interaction, and its consequences to individual buildings. While the real life city model including the drainage system has been provided by the authorities of the city of Munich and comprises an area of about 2 by 2 km with detailed topography and a complete set of approximately 3,000 buildings modelled on LOD 1 of CityGML, IFC-models are the initial starting points for generating octrees on the level of individual buildings. In order to investigate the effects of the drainage system collapsing due to a heavy rain scenario, a fully three-dimensional parallel free surface flow simulation is incorporated with the interaction of the one-dimensional pipe-network flow. The whole simulation is performed on three levels of resolution, each of which is discretised by a fast voxelisation algorithm to generate a computational grid for the CFD simulation.

V. Varduhn (✉)
Department of Civil Engineering, University of Minnesota, 500 Pillsbury Drive S.E., Minneapolis, MN 55455, USA
e-mail: vvarduhn@umn.edu

R.-P. Mundani · E. Rank
Computation in Engineering, Technische Universität München, Arcisstraße 21, 80333 Munich, Germany
e-mail: mundani@tum.de

E. Rank
e-mail: ernst.rank@tum.de

© Springer International Publishing Switzerland 2015

M. Breunig et al. (eds.), *3D Geoinformation Science*,
Lecture Notes in Geoinformation and Cartography,
DOI 10.1007/978-3-319-12181-9_4

# 1 Introduction

Many important scientific and technological developments in the last few decades have resulted in a tremendous increase in availability of 3D geometric models on different scales. To name a few, Google Earth is a many-scale model with resolutions from continental down to building scale, 3D Geographic Information Systems (GIS) (Yeung and Lo 2002; Bolstad 2005) such as CityGML (Kolbe et al. 2005; Kolbe 2007) are available for entire cities and regions, Building Information Models (BIM)[1] (SCRA 2006; Pazlar and Turk 2008) are starting to revolutionise planning and construction of buildings and infrastructure. All these models can be made available via the internet always and everywhere. Yet, whereas GIS and BIM are very well established as complex information resources and widely used for various planning aspects, they are much less established as a basis for advanced numerical simulations. On the other hand, tremendous progress has been observed in computational sciences and engineering over the past decades, making it now possible to perform large scale computations based on geometric models spanning many spatial and also temporal scales. Earthquake simulation is one area with spectacular advances in recent years, and material scientists start to understand structural behaviour by micro-macroscopic simulation based on first geometric and physical principles. In this paper, we will present recent results on coupling geographic, infrastructure, and building models to multi-resolution numerical simulations. In order to achieve this, a parallel data access framework with interfaces to all parts of the simulation pipeline such as pre-processing, numerical simulation, and post-processing has been developed.

Our discussion will first focus on suitable distributed data structures to store these models, building on forests of space-tree models. On a global level, an octree is defined, yielding efficient access to all objects in a city model. Leafs of this octree contain pointers to at most one building or infrastructure object. These local objects with all their components and construction elements are stored themselves using space-trees with access not only to geometric but also to semantic information via the underlying building information model.

From this hierarchical data model the framework now allows to derive all necessary input for a multi-resolution numerical simulation. We will concentrate in this paper on computational fluid dynamics (CFD), noting that other types of simulation such as traffic flow or evacuation scenarios could be addressed by the same conceptual approach. Starting from the large-scale model, a grid of voxels is derived. For realistic simulations these grids tend to be huge, and demand for specialised algorithms which take advantage of high-performance computing. Simulation results from this global model are now projected to initial and boundary conditions for simulations on local computational models which are, depending on the desired resolution, derived from the suitable levels of the multi-resolution geometric models.

---

[1] BuildingSMART. http://www.buildingsmart-tech.org/speci.cations/ifc-releases/ifc4-release.

The applicability of the complete approach presented in this work is shown by simulating urban flooding including surface flow of a city, the pipe network interaction, and its consequences to individual buildings. While the real life city model including the drainage system has been provided by the authorities of the City of Munich and comprises an area of about 2 by 2 km with detailed topography and a complete set of approximately 3,000 buildings modelled on LOD 1 of CityGML, IFC-models are the initial starting points for generating octrees on the level of individual buildings. In order to investigate the effects of the drainage system collapsing due to a heavy rain scenario, a fully three-dimensional parallel free surface flow simulation is incorporated with the interaction of the one-dimensional pipe-network flow. The whole simulation is performed on three levels of resolution, each of which is discretised by a fast voxelisation algorithm to generate a computational grid for the CFD simulation.

## 2 Framework

One contribution of this work focuses on the development of a framework for handling large amounts of complex data, providing efficient access and evaluation strategies to retrieve and process these data in real-time, including fully detailed product model data of constructions and built infrastructure, large scale GIS data such as highly resolved terrain descriptions, and the definition of pipe networks. Furthermore, a city model definition is given which embeds individual buildings in the urban context, see Figs. 1 and 2 for examples representing different scales.

In order to efficiently organise and orchestrate those different data in one integrating structure, a hierarchical approach is used. Therefore, a so-called *first level octree* is generated based solely on the bounding boxes of all product models.

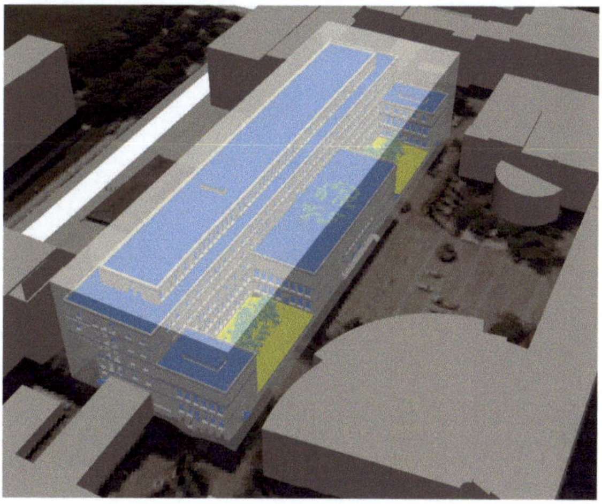

**Fig. 1** City model (showing Technische Universität München's main campus) consisting of terrain, GIS, and highly detailed BIM data

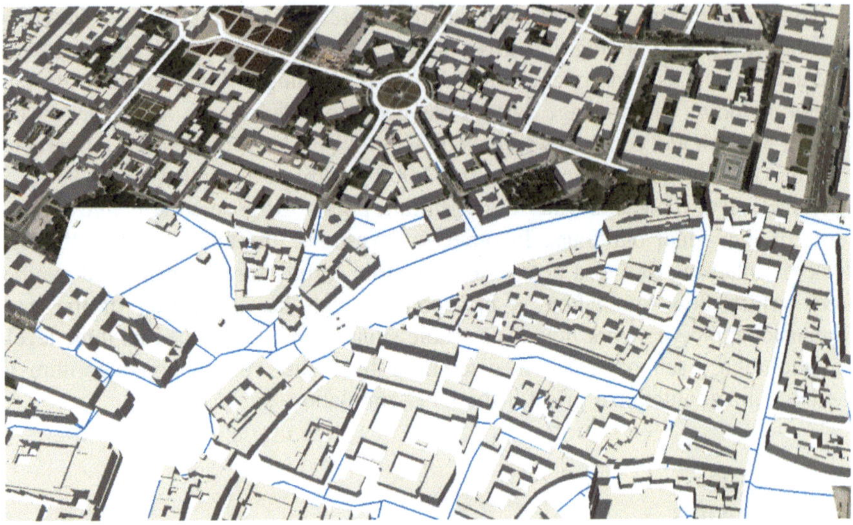

**Fig. 2** City model (showing Munich City Centre) consisting of terrain and GIS data including the underground water drainage system

Instead of storing the model data itself, the voxels contain just a link to the corresponding resource, i.e. the location of the respective file containing the real product model data, see Fig. 3.

As this first level octree contains only links to product models instead of real data, a nearly unlimited number of constructions and buildings can be efficiently orchestrated by this approach even in real-time. Any access to specific product model data is possible at all times, while questions concerning range or proximity of buildings can be answered directly from the first level octree. Nevertheless, to process several product model data in real-time, a parallel concept has been developed. A master process assigns different product model data to its slave

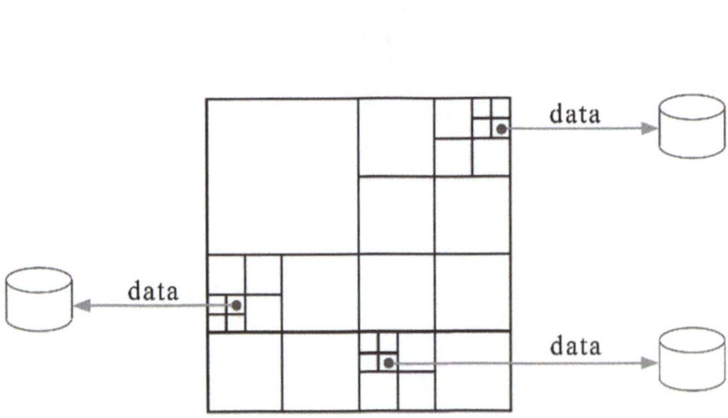


**Fig. 3** First level octree of an entire region or city storing/providing links to the corresponding product model (IFC) or CityGML data

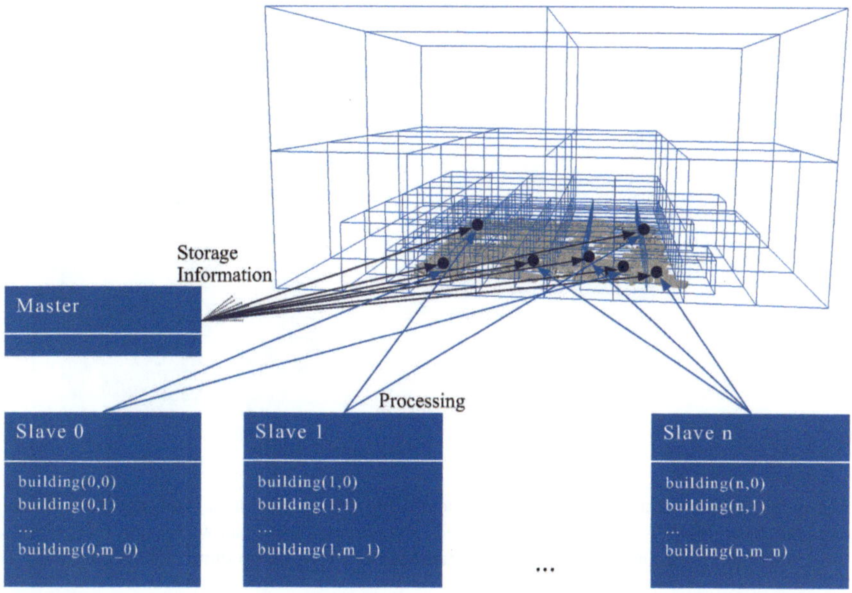


**Fig. 4** Parallel processing of product model data via a distributed memory approach (MPI) on the global scale—one master process handling the first level octree assigns product model data to its slave processes for further treatment

processes for further treatment, i.e. for either the visual exploration of these data or the derivation of computational meshes as input for subsequent simulation tasks. Figure 4 shows the principle of parallel processing of several buildings embedded into the first level octree.

The processing of product model data follows the same octree-based approach and is thus called *second level octree.* An octree is now generated from a single IFC model, representing more or less details depending on the chosen resolution, i.e. tree depth $d$, leading to $2^d \times 2^d \times 2^d$ voxels in case of a fully refined tree.

The second level octrees—in contrast to the first level one—are not persistently stored and solely computed on-the-fly (i.e. demand-driven) subject to proximity reasons for (interactive) visual exploration purposes or level-of-detail considerations for subsequent simulation tasks. For any voxel of the second level octree it can be decided at all times which parts of the IFC model are mapped, i.e. which parts (or semantic entities) of a building are intersected or included by the respective voxel, using unique part IDs (inherited by all of a part's faces describing its geometry) stored along with the voxel data. Thus, there is a strong linkage between the second level octree and the product model, necessary to answer questions concerning insight gained by the numerical simulation and its impact on entities described in the GIS or BIM model.

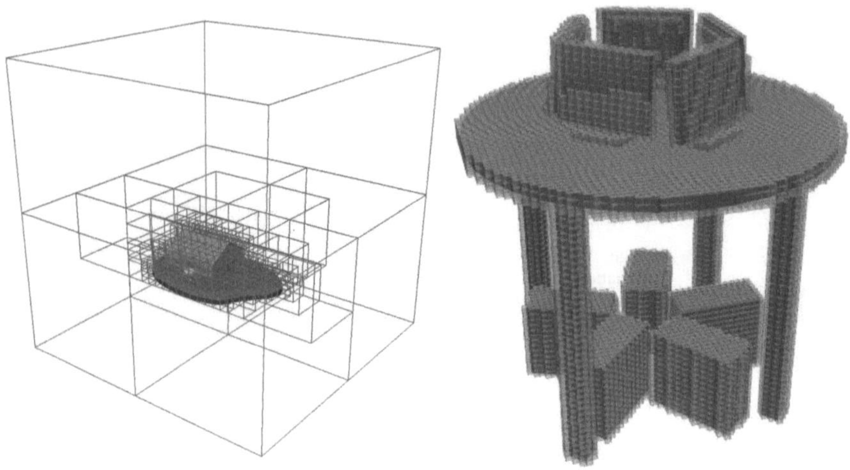

**Fig. 5** Examples of a second level octree for a building information model with depth 5 consisting of 84,000 triangles (*left-hand side*) and a derived computational mesh (shown are only cells with flag '1') of auxiliary information located inside a building for a local-scale simulation (*right-hand side*)

As the focus of our approach lies on coupling multi-resolution geometry to environmental numerical simulations, computational meshes of different resolutions representing different subsets of the entire GIS/BIM domain have to be generated. Therefore, users can define a region of interest (cf. bounding box) to select parts of the scene (whole regions, several buildings, or just few rooms of one building) for which a computational mesh shall be derived. The methodological concept is as follows. For each involved IFC model (decided by intersecting the region of interest with the first level octree) a second level octree up to a certain resolution is generated. From these octrees regular computational meshes are derived by further uniform refinements, constituting the entire computational domain. For each cell this computational domain stores its corresponding flag, i.e. '1' or '0', indicating a cell belonging to the geometry ('1') or to the fluid part ('0'). Figure 5 shows a second level octree together with a computational mesh for some sample IFC model. Again we would like to mention, that for any voxel of the second level octree and, thus, also for any cell of the computational mesh it can be retrieved to which (semantic) part of the IFC model it maps, hence an evaluation of simulation results on the product model is possible in order to assess which parts are affected, e.g. by urban flooding, as in our sample application.

As the computational meshes serving as input to numerical simulations can easily exceed several millions or even billions of cells, appropriate sparse storage schemes such as (Pissanetzky 1984; Saad 1992) are of utmost importance. In such schemes, only values different from '0' are stored in a vector together with their respective row and column index of every row sequence of the data.

## 3 Multi-resolution Parallel Numerical Simulation

During the past years, multi-scale or multi-resolution approaches have frequently been applied in fluid flow applications, see (Grinberg et al. 2011; John and Kindl 2010; Zhao et al. 2004) for instance. Having high-resolution GIS/BIM model data at hand, the novelty of our approach lies in directly coupling these models to fluid flow simulations at different scales (city → building → room, e.g.) and to propagate simulation results back to those models, e.g. for the assessment of possible damage as impact of an urban flooding. As our framework keeps the link between product model data and the corresponding computational grids, any mapping of simulation results to geometric and semantic information of IFC models is easily possible, in order to answer questions such as "*which parts of a building were exposed to flooding*" or "*which electric devices were damaged due to contact with water*".

The simulation of three-dimensional free surface flow is performed by utilising the parallel OpenFOAM solver *interFoam* on some cluster or supercomputer. In order to perform a flood simulation, the three-dimensional incompressible Navier-Stokes equations are applied (Ferziger and Peric 1996). These equations are solved on a regular Cartesian grid (derived from the second level octree), following a Finite Volume (FV) approach. The interface between the gas-phase region and the liquid-phase region is tracked by a Volume-of-Fluid (VoF) concept (Deshpande et al. 2012).

Resolution adaptation increases the accuracy of the numerical simulation by performing a sequence of successively refined simulations, where the results on the coarse scale are projected as initial and boundary condition to the fine scale. Therefore, the simulation on the global domain (typically covering the entire region) is stopped at a predefined time. Then coarse scale simulation values are used as initial and boundary condition for the refined simulation on the local domain (being a subset of the global one, i.e. typically covering few or just single buildings). Special focus has to be put on the correct application of (transient) boundary conditions, as investigated in Varduhn and Parallel (2014).

## 4 Post-processing—Propagation to Product Model Data

After performing a flow simulation, the quantities of interest (water-level, velocities, pressure, e.g.) are available for all chosen levels of detail. Algorithms for graphical investigation of these results are well known from scientific visualisation, such as isosurfaces, streamlines, slice planes, etc. (Hansen and Johnson 2005).

Due to the linkage (GIS/BIM model data ↔ simulation results) over different scales, a very detailed investigation as shown in Fig. 6 is now possible. Quantities of interest obtained from the numerical simulation can be used e.g. for a protection planning, mapping individual results to relevant entities of buildings, equipment or built infrastructure. Questions concerning damage probability and protection

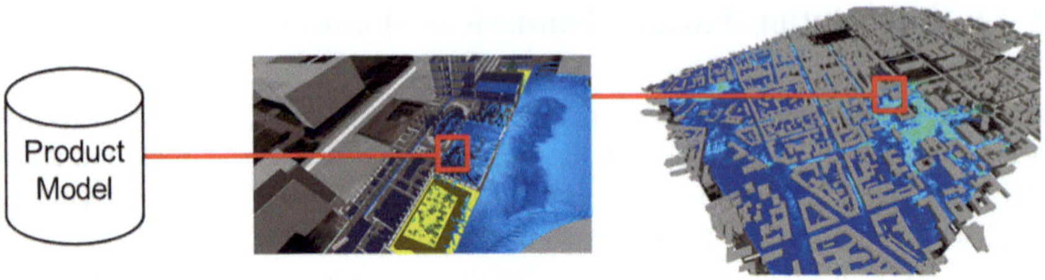


**Fig. 6** Linkage between product model data and flow simulation results on different scales—local scale (*middle*) and global scale (*right*)—in order to explore possible flood impacts

possibilities for an assumed flood can be answered in a rational, data and simulation supported disaster management.

The key for this insight is the relation between the second level octree (generated from product model data) and the derived computational mesh as described in Sect. 2. As depicted in Fig. 7, the fluid domain is connected to the structural representation of the product model data, mapping boundary faces of the computational mesh to the construction entities of the underlying product model data. This link is perpetuated throughout the hierarchical approach presented.

Having the results of the flow simulation at hand, simulated quantities are available over the boundary of the computational domain. For CFD simulations, these quantities contain the fluid velocities, the hydrostatic pressure, and water-level. In order to evaluate now the impact of flooding, the following steps have to be performed. If a user wants to estimate the danger for damage of a specific part—for instance some expensive facilities—he can use that part's ID to query for the

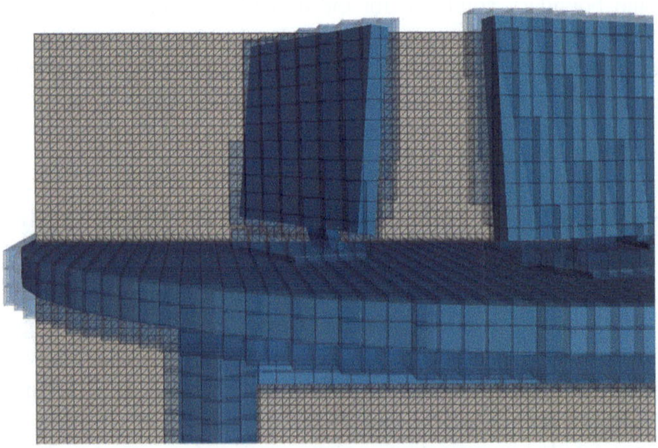

**Fig. 7** The computational mesh of the fluid flow simulation can be mapped to the construction details of the product model data. The boundary faces of the fluid mesh are shared with the octree representation of the construction detail. This mapping gives the linkage between simulation results and construction details affected

respective voxels of the second level octree and, thus, the associated computational mesh. Within this mesh, the fluid domain and the voxelisation of that part's faces (triangles) share a common boundary, on which the computed quantities can be evaluated. Thus, the impact of the water to that part can be determined (for instance the expected water-level), and the endangerment can be estimated based on the rich underlying semantics.

In case entities of interest do not exhibit an explicit geometric representation, such as the semantic model 'room' consisting of its surrounding IFC parts wall, floor, and ceiling, those relations have to be established first (see Hitchcock and Wong 2011, e.g.), before being subject to the proposed flood assessments.

## 5 Urban Flood Simulation with Pipe Network Interaction

The European floods in 2002 and 2013 are examples of devastating impacts on regions, cities, buildings, constructions, and—first and foremost—people affected by them. In May 2013, the Institution of Civil Engineers (ICE) held the ICE Flooding 2013 (Balmforth and Benyon 2013) in order to bring together scientists and researchers to develop flood resilient communities. It is now common understanding that extensive flood simulation will play an important role to achieve this goal.

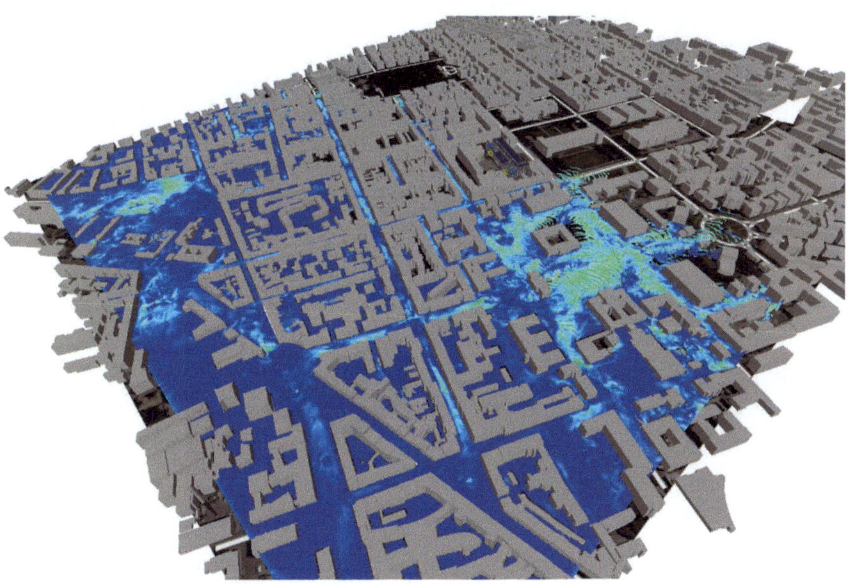

**Fig. 8** Flood simulation based on a detailed city model of Munich, visualised using *colour encoded* (velocity) contour surfaces

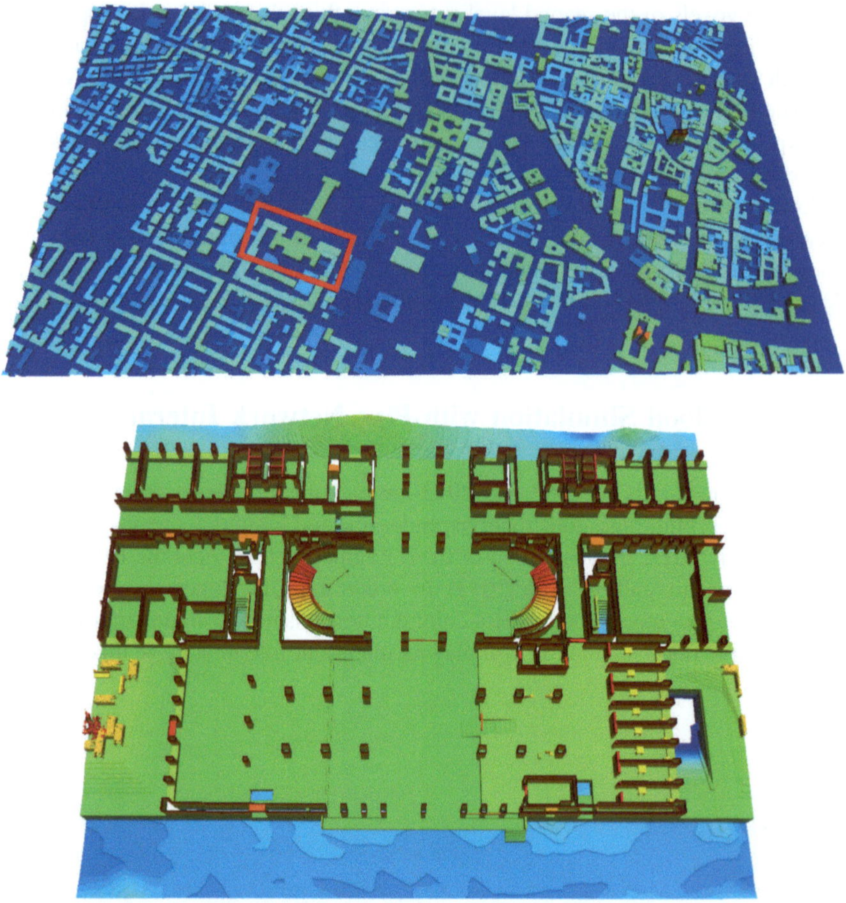

**Fig. 9** Computational domain depicted by a contour plot for the global (*top*) and the local scale (*bottom*). The relative height of the surface is *colour encoded*

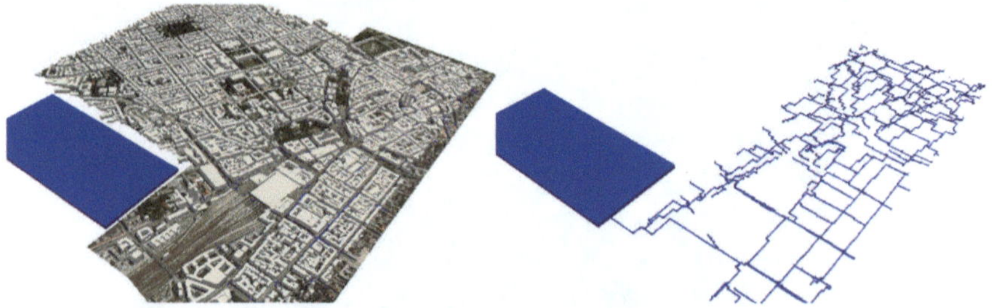

**Fig. 10** Modelling a collapsing drainage water system by placing constant hydrostatic pressure (from a fictitious large water reservoir) on one inlet of the network. Initially, the pipe network is assumed to be already flooded

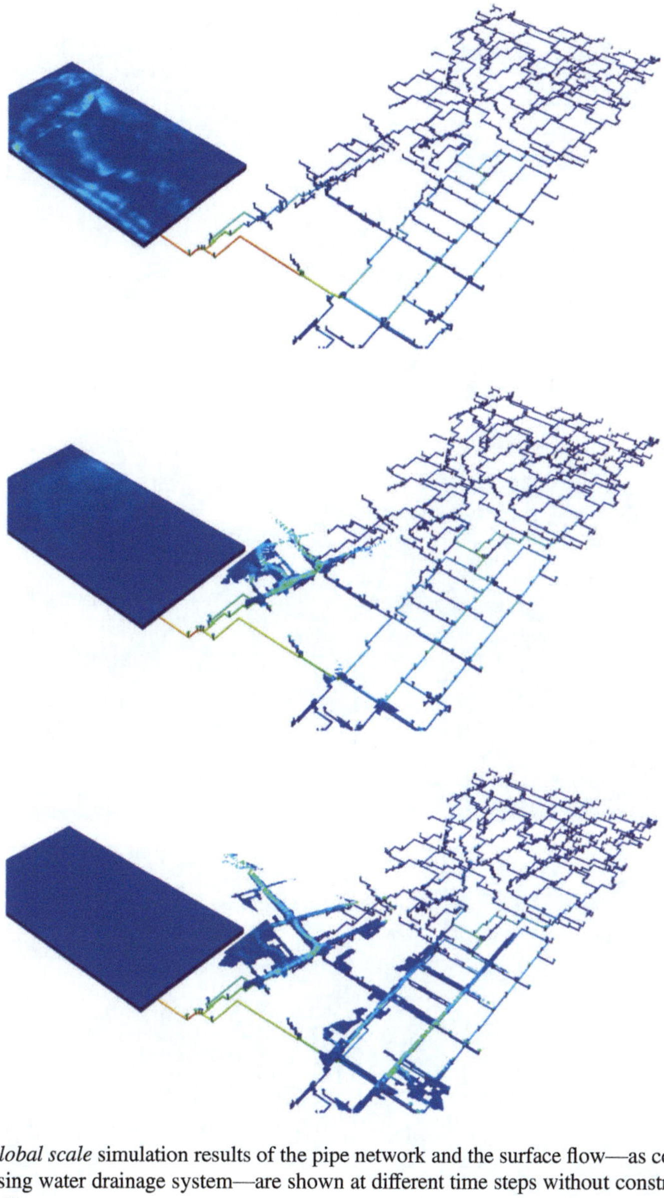

**Fig. 11** *Global scale* simulation results of the pipe network and the surface flow—as consequence of a collapsing water drainage system—are shown at different time steps without constructions for better visibility

In the present work we study the interaction of an overloaded pipe network in a city's drainage system and the surface flow following an assumed heavy rainfall. Figure 8 shows the results of such an urban flood simulations.

For this test scenario, a certain region of the Munich City Centre is affected by heavy rainfall, which causes hydrostatic pressure on parts of the rain water drainage

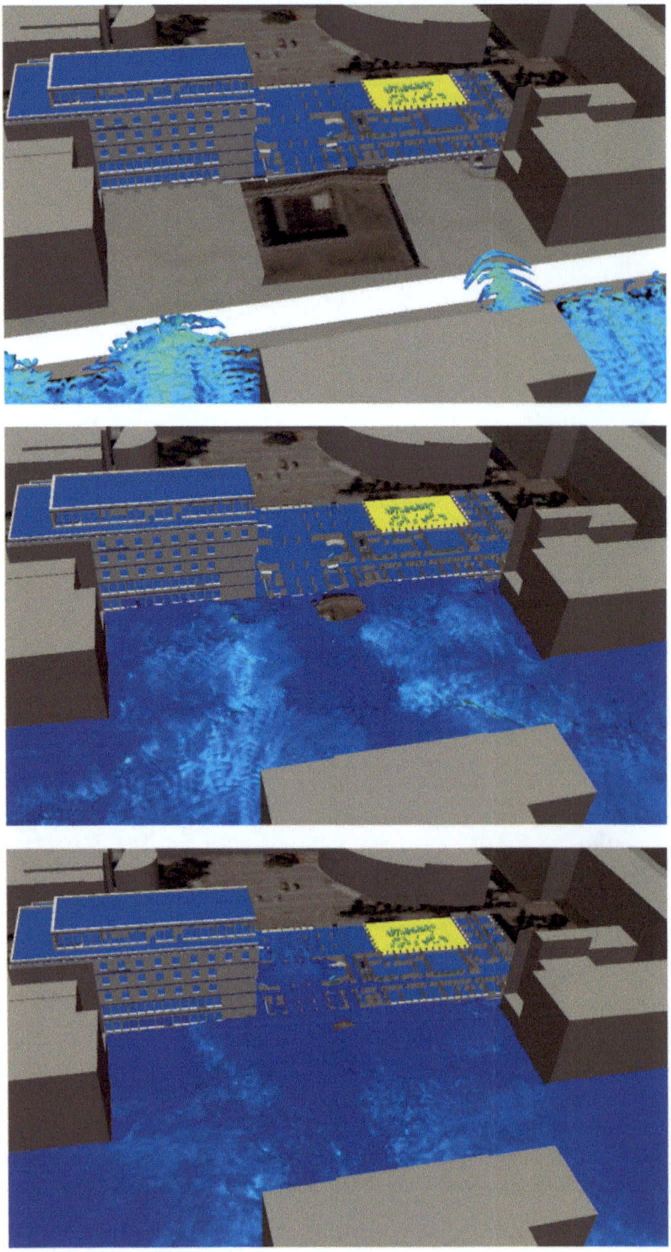

**Fig. 12** *Medium scale* investigation of flow behaviour at different time steps on single buildings after first resolution refinement

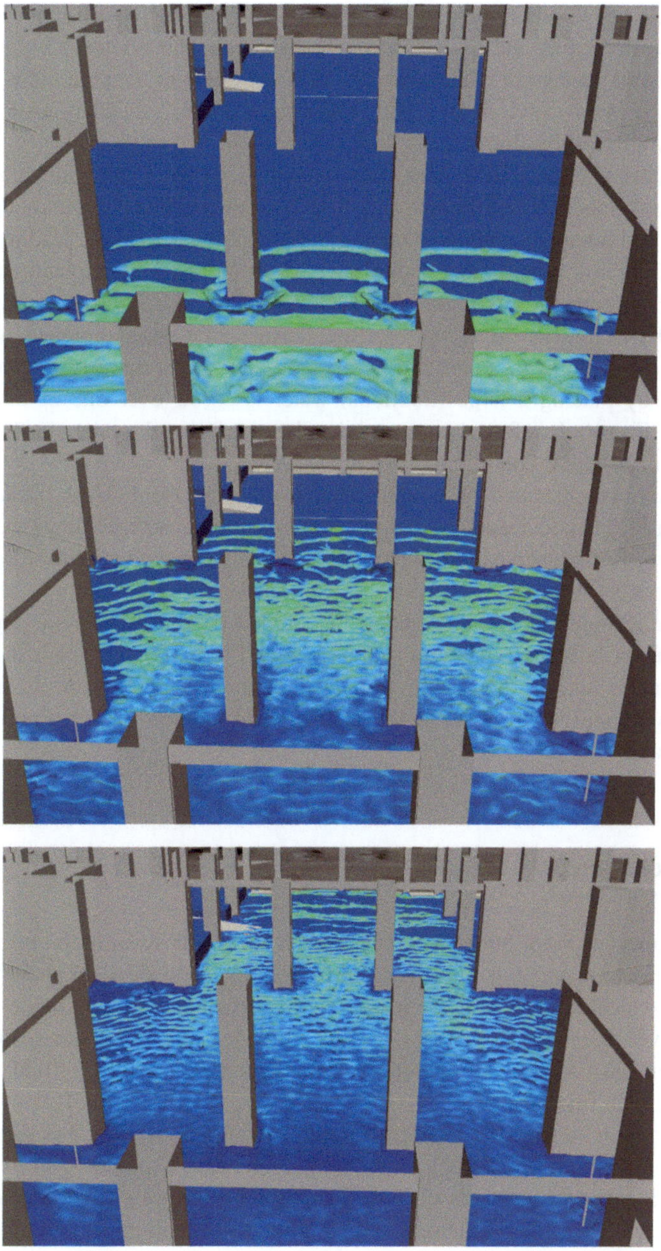

**Fig. 13** *Local scale* investigation of flow behaviour at different time steps on construction details after second resolution refinement

system and the sewers in that area. Due to this hydrostatic pressure, the flow in the pipe network evolves and spills parts of the city, even where no or only little rainfall was observed. Using our framework, the flow evolution in the pipe network and on the surface can be investigated, thus allowing a detailed evaluation of areas and buildings affected by floods.

The used GIS and BIM data covers Munich's City Centre including terrain descriptions, models of buildings, and the pipe network. The computational domain is discretised with a resolution of up to $1{,}800 \times 1{,}650 \times 75$ voxels, leading to a total of 222.75 million cells. Figure 9 shows a contour plot of the computational domain (boundaries) both for the global and the local scale.

In order to impose the scenario of a collapsing rain water drainage system, the sewer network is given a constant hydrostatic pressure at one of its inlets. This is achieved by placing a large (fictitious) water reservoir over one of the inlets as shown in Fig. 10. It is further assumed that the pipe network system itself is already flooded as an initial condition.

The simulations have been conducted using the OpenFOAM package *interFoam*. In Fig. 11, the results of the pipe network and the surface flow (global scale) are shown. For better visibility, the constructions were removed in order to study flood impacts on the city.

Figure 12 highlights the first adapted resolution refinement for studying the flow behaviour on buildings (medium scale). In Fig. 13, the second adapted resolution refinement is shown, here with projected medium scale results as initial and boundary conditions in order to investigate flood impacts on construction details (local scale).

## 6 Conclusions

In this paper, we have presented a mashup for the synthesis of high-resolution GIS and BIM data to serve as input for multi-resolution numerical simulations. Due to the hierarchical organisation of the underlying terrain and product model data with a two-level octree-based concept, we are not only able to handle such complex scenes but also to derive computational meshes for subsequent simulation tasks such as urban flooding.

Furthermore, due to the linkage between product model data and the computational domain we can fed back simulation results to IFC models and evaluate the computed quantities on parts of those models. This allows us to estimate possible damage caused e.g. by flood impacts on buildings and built infrastructure in an easy and efficient way. Hence, the shown methodologies can serve and support planers in flood assessments and protection planning. With the extension of pipe networks to urban flood simulations, we have also shown the effective integration of additional data to such multi-resolution computations.

## References

Balmforth D, Benyon R (2013) Developing flood resilient communities. In: Proceedings of the ICE flooding 2013

Bolstad P (2005) GIS fundamentals: a first text on geographic information systems. Eider Press, St. Paul

Deshpande SS, Anumolu L, Trujillo MF (2012) Evaluating the performance of the two-phase flow solver interFoam. Comput Sci Discov 5(1):014016

Ferziger JH, Peric M (1996) Computational methods for fluid dynamics, vol 3. Springer, Berlin

Grinberg L, Insley JA, Morozov V, Papka ME, Karniadakis GE, Fedosov D, Kumaran K (2011) A new computational paradigm in multiscale simulations: application to brain blood flow. In: Proceedings of the international conference for high performance computing, networking, storage and analysis. IEEE

Hansen CD, Johnson CR (2005) The visualization handbook. Elsevier, Amsterdam

Hitchcock RJ, Wong J (2011) Transforming IFC architectural view BIMs for energy simulation. In: Proceedings of building simulation

John V, Kindl A (2010) A variational multiscale method for turbulent flow simulation with adaptive large scale space. Comput Phys 229(2):301–312

Kolbe TH (2007) CityGML–3D geospatial and semantic modelling of urban structures. Presentation on the GITA/OGC emerging technologies summit in Washington

Kolbe TH, Gröger G, Plümer G (2005) CityGML—interoperable access to 3D city models. In: Proceedings of the first international symposium on geo-information for disaster management

Pazlar T, Turk Z (2008) Interoperability in practice: geometric data exchange using the IFC standard. ITcon—Special Issue case Studies BIM use 13:362–380

Pissanetzky S (1984) Sparse matrix technology. Academic Press, London

Saad Y (1992) Numerical methods for large eigenvalue problems, vol 158 SIAM

SCRA (2006) STEP application handbook: ISO 10303

Varduhn V (2014) A parallel, multi-resolution framework for handling large sets of complex data, from exploration and visualisation to simulation. Ph.D. thesis, TU München

Yeung AKW, Lo CP (2002) Concepts and techniques of geographic information systems. Upper Saddle River, NJ

Zhao Q, Armfield S, Tanimoto K (2004) Numerical simulation of breaking waves by a multi-scale turbulence model. Coast Eng 51(1):53–80

# Crisp Clustering Algorithm for 3D Geospatial Vector Data Quantization

**Suhaibah Azri, François Anton, Uznir Ujang, Darka Mioc and Alias A. Rahman**

**Abstract** In the next few years, 3D data is expected to be an intrinsic part of geospatial data. However, issues on 3D spatial data management are still in the research stage. One of the issues is performance deterioration during 3D data retrieval. Thus, a practical 3D index structure is required for efficient data constellation. Due to its reputation and simplicity, R-Tree has been received increasing attention for 3D geospatial database management. However, the transition of its structure from 2D to 3D had caused a serious overlapping among nodes. Overlapping nodes also occur during splitting operation of the overflown node $N$ of $M + 1$ entry. Splitting operation is the most critical process of 3D R-Tree. The produced tree should satisfy the condition of minimal overlap and minimal volume coverage in addition with preserving a minimal tree height. Based on these concerns, in this paper, we proposed a crisp clustering algorithm for the construction of a 3D R-Tree. Several datasets are tested using the proposed method and the percentage of the overlapping parallelepipeds and volume coverage are computed and compared with the original R-Tree and other practical approaches. The experiments demonstrated in this research substantiated that the proposed crisp clustering is capable to preserve minimal overlap, coverage and tree height, which is advantageous for 3D geospatial data implementations. Another advantage of this approach

S. Azri (✉) · U. Ujang · A.A. Rahman
3D GIS Research Group, Department of Geoinformation, Faculty of Geoinformation and Real Estate, Universiti Teknologi Malaysia, 81310 Skudai, Johor, Malaysia
e-mail: norsuhaibah@gmail.com

U. Ujang
e-mail: mduznir@utm.my

A.A. Rahman
e-mail: alias@utm.my

F. Anton · D. Mioc
Department of Geodesy, National Space Institute, Technical University of Denmark, Elektrovej 328, 2800 Kongens Lyngby, Denmark
e-mail: fa@space.dtu.dk

D. Mioc
e-mail: mioc@space.dtu.dk

© Springer International Publishing Switzerland 2015

M. Breunig et al. (eds.), *3D Geoinformation Science*,
Lecture Notes in Geoinformation and Cartography,
DOI 10.1007/978-3-319-12181-9_5

is that the properties of this crisp clustering algorithm are analogous to the original R-Tree splitting procedure, which makes the implementation of this approach straightforward.

**Keywords** 3D spatial data management · 3D spatial data clustering · 3D Geo-DBMS · 3D spatial indexing

## 1 Introduction

The uses of three-dimensional (3D) data input, 3D analysis and 3D visualization in geospatial applications, that are designed for managing and organizing geographic data (i.e., such as surface and subsurface objects), have recently become more widespread. This is due to their capability in handling efficiently 3D spatial information with non-spatial attributes. At the same time, various spatial problems could be solved efficiently (such as heat and sunlight spreading, calculation of wave propagation and noise spreading prediction models) in 3D rather than in 2D. 3D applications could be seen in various disciplines such as navigation, city planning and management, natural risks prevention, physical modeling and crime analysis.

The trend of 3D geospatial data is influenced by two main factors: users demand and data acquisition technology development such as Light Detection and Ranging (LiDAR), Interferometric Synthetic Aperture Radar (InSAR), and Unmanned Aerial Vehicles (UAV). Users demand on 3D geospatial application is due to the realistic view of environment which allows users to easily comprehend the real world problems straightforwardly (Izham et al. 2010; Jazayeri 2012; Uznir et al. 2013b). With the aid of data acquisition technologies, acquiring 3D data is getting much more convenient for 3D geospatial applications. Users could have various choices of data formats and resolutions depending on the tools and techniques used. On the other hand, the size of these 3D datasets is commonly very large due to the geometric details and in certain cases, it is incorporated with semantic datasets. For instance, by using laser scanning devices, millions of 3D points are captured for a single building or object. The same issue could be seen in 3D data interpretation applications such as CityGML. In CityGML environments, hundreds or thousands of coordinates and information are coded in XML format and stored for each urban object. In addition, in Building Information Modeling (BIM) applications, detailed building geometries and their descriptions are stored for each building model and at the same time, it escalates the data storage size.

Due to the large sizes of detailed information within 3D geospatial data, an efficient approach of 3D data management is needed. The management of 3D geospatial data is important with the aim to ensure the 3D data is efficiently managed and the produced information is efficiently retrieved. Due to these challenges, experts and spatial professionals are still working with the management of 3D geospatial information for geospatial applications. Therefore, in this paper, a method for improving the management of 3D geospatial information is presented. This method is useful for improving the efficiency of spatial access methods in order to manage clustering and indexing of 3D spatial datasets and reduce processing time during data retrieval process.

This paper is organized as follows. Problems and motivations regarding 3D geospatial information management are reviewed in the next section. In Sect. 3, the concept of the proposed indexing method is explained with its implementation. In Sect. 4, we present the analysis and results for the experiment conducted in this paper. Finally, in Sect. 5, we present the conclusions of this research.

## 2 Research Problem and Motivation

The increasing number of 3D geospatial applications could be seen in various fields and areas such as urban planning and management, natural risks modeling and prevention, cadaster and land administration, terrorism modeling and prevention. In the next few years, it is expected that 3D will be an intrinsic part of the core geospatial data infrastructure rather than a distinctive add-on as it is now. However, there are several issues that need to be resolved before utilizing 3D as a core geospatial data infrastructure. One of the challenges is the optimal management of 3D geospatial data, with optimization functions that combine several factors such as disk space consumption, data model, data structure and spatial access method. These factors are related to one another in order to achieve and optimize the efficiency of data management. However, 3D data come with a costly storage size, due to detailed geometrical features and detailed non-spatial descriptive information as discussed in the previous section. For an example the size of 3D geospatial data produced from laser scanning techniques, an interactive editing of large point clouds for a number of synthetic and real-world examples would take up to 63 GB (Wand et al. 2007). Another example of storage size for 3D geospatial data could be seen in (Uznir et al. 2013a). From the test in (Uznir et al. 2013a), 10,084 kb of disk storage is required for 2,500 buildings in CityGML.

Information carried and attached with 3D data varies and is amalgamated with other useful information such as the object's geometry (e.g., line, point and polygon) and other semantically based information such as roof, wall, color, object's id, material, etc. These geospatial data and information are very complex to handle in spatial database management systems. Thus, an enhanced spatial data model with spatial indexing methods is employed in order to manage the complexities and the large quantities of geospatial information. Based on this data model, 3D data are interpreted and transformed into a set of records or spreadsheets. To boost up the added value of these records, a 3D geospatial database is utilized for an efficient data constellation. In this platform, records will be stored, analyzed and manipulated efficiently. However, the efficiency of this platform towards a massive 3D dataset is slightly affected due to data complexity and disk storage consumption. In conjunction to this issue, the performance of data retrieval is also affected and decreased. The time of data retrieval is important in most critical applications, especially for real time applications or service based applications such as emergency response centers which require the retrieval of precise information at the right time in the fastest possible way.

In order to boost up the efficiency of data retrieval, a spatial index structure is utilized together with query operators. By using an index structure, the search area is minimized within a certain range without having a thorough inspection for the whole records. By implementing this approach, processing time is reduced as well as information retrieval time. For real time applications, spatial index structures are best to be utilized with data visualization and data updating (Liu et al. 2010; Zhu et al. 2007). However, in most of commercial spatial database management systems, supported index structures are mainly 2D. 2D index structures are not the best fit solution to be used for 3D geospatial information since the data types and relationships between objects are defined differently than in 2D. Until now, a well-established index structure for 3D spatial information is still an open research problem. Thus, a dedicated index structure for 3D geospatial information is significant for efficient data analysis and data retrieval.

There are various types of spatial indices invented for 2D application such as *k*-d-tree and Quad-tree. Among these structures, the R-Tree structure reported in Guttman (1984) is the most widely used index in spatial databases. This could be seen from the increasing number of R-Tree variants since it was invented in 1984 in Manolopoulos et al. (2006, Balasubramanian and Sugumaran (2012). Due to the reputation and the simplicity of its structure, many researchers (Zlatanova 2000; Arens et al. 2005; Zhu et al. 2007; Wang et al. 2010) agreed that the transition of R-Tree from 2D to 3D would be a starting point towards a promising 3D spatial index structure.

The implementation of the 3D R-Tree at geospatial database level was initiated by several commercial database management system producers, such as Oracle Spatial. Since then, several issues have been addressed due to the dimensionality expansion from 2D to 3D. In general, the features and properties of R-Trees would be inherited by 3D R-Trees, since the only modification is its dimensionality expansion. However, the dimensionality expansion from 2D to 3D did not address all the issues on overlapping nodes and splitting operations. According to Zhu et al. (2007), the parallelepipeds in 3D R-Trees will frequently overlap one another and the parallelepipeds of nodes could even contain one another. In R-Trees, overlapping among nodes should be minimized to avoid redundancy or replicated entries in a different node.

Another issue of R-Tree construction is a node splitting operation for overflown node $N$ with $M + 1$ entries. According to Fu and Teng (2002), Sleit (2008), Liu et al. (2009), Korotkov (2012), the splitting operation is the most critical process of R-Trees, since tree structure is altered in this phase. Each overflowing node $N$ needs to be split into two nodes and at the same time, it should produce minimal tree height, reducing overlapping area among rectangles and minimal coverage area of the node. These issues become critical when it comes to 3D. The minimization of overlap coverage of parallelepipeds is more complex and the splitting operation requires a different approach than in 2D.

The classical R-Tree structure (Guttman 1984) proposed three different approaches of splitting method i.e. quadratic, linear and exhaustive. However, these approaches have been optimized from time to time by scientists and researchers in order to produce minimum non-overlapping nodes (Paar 2006; Zhu et al. 2007; Sugihara and Shen 2012; Xu and Coors 2012).

# 3 Crisp Clustering Algorithm for 3D R-Tree

## *3.1 Splitting Algorithm for R-Tree Construction*

The splitting operation is the most crucial process during the R-Tree construction. At this phase, the tree structure is altered either at the child or parent level or both. Splitting happens when the node *N* holds more than M entries, where M is the maximum number of entries for each node. Each node *N* is only allowed to hold $n \leq M$ entries, where *n* is number of entries in *N*. As shown in Fig. 1, the parallelepiped *P* denotes the overflown node *N* that requires a splitting operation with the maximum number of entries $M = 4$.

The original splitting approach in Guttman (1984) suggests to apply either exhaustive, quadratic or linear splitting algorithm. The exhaustive approach splits the overflown node by exhaustively visiting all the possible groupings. Then, the best group is chosen with respect to the smallest coverage area. However, based on the reviews from Manolopoulos et al. (2006), Al-Badarneh et al. (2013), Sleit and Al-Nsour (2014), the exhaustive approach is not practical due to the exponential time required.

Meanwhile for the quadratic approach, two entries will be selected to initiate two new nodes. Then, the remaining entries are selected one by one into the new nodes. The entry is added to the node whose coverage is minimized by the addition to the

**Fig. 1** Parallelepiped *P* with an overflown node *N*

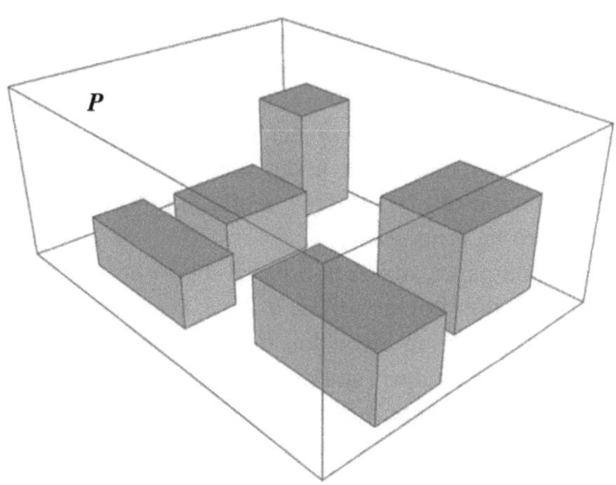

new nodes. All remaining entries will go through this process until one of the new nodes reached the maximum number of entries ($M - m + 1$). According to Guttman, the quadratic approach makes compromises to achieve reasonable retrieval performance. However it does not guarantee to find the minimal or smallest coverage area compared with exhaustive approach.

The linear split algorithm has a linear time complexity. Two objects that are separated farthest apart are chosen as new nodes. Then remaining objects will be assigned one by one to the nearest node. All of these suggested methods are proven to work correctly for the construction of a R-Tree structure. However, since the main concern of R-Trees is to have very minimal areas of overlapping nodes, several researches have been conducted to improve the original methods. For example, the new linear algorithm in Ang and Tan (1997) is the improved version of Guttman's linear algorithm. In order to minimize overlap between nodes, this approach calculated the distributions of objects along all main axes to produce farthest distance towards the boundary of the overflown node. Another approach of splitting method is proposed in Kamel and Faloutsos (1994). In their approach (Kamel and Faloutsos 1994), they use the technique of space filling curves for nodes ordering and grouping all the nodes to minimize the area and perimeter of the minimum bounding rectangle.

## 3.2 Crisp Clustering Partition Based Algorithm for 3D R-Tree

Splitting and clustering are two different topics. However, the possibility of merging these two different topics within the construction of 3D R-Trees seems possible and obvious. The main idea of this paper is to cluster 3D geospatial objects based on clustering classes. Objects are then split according to their classes. One of the possible clustering methods to be utilized in this case is the crisp clustering algorithm. The crisp clustering algorithm has been used ubiquitously from web mining, spatial data analysis, business, prediction based on groups and many more. This approach considers non-overlapping partitions and each data vector either belongs to a class or not. This characteristic is suitable with the 3D R-Tree condition which is that an object will be clustered and appear only once in an index node so that the index nodes and their total number will be minimized. In the construction of a 3D R-Tree, replicated objects in several nodes will cause a larger structure than necessary.

In this paper, a partition based algorithm for crisp clustering will be utilized to construct the structure of 3D R-Tree. *k*-means (MacQueen 1967), is one of the most popular clustering algorithm used to solve various fundamental problems in data mining, data analysis and machine learning. The *k*-means algorithm is a simple and fast clustering algorithm although it offers no approximation guarantee at all. The *k*-means algorithm for a set of 3D vector data is described as follows:

| | |
|---|---|
| Input | A set of 3D vector data $P = \{p_1, p_2, \ldots, p_n\} \in \mathbb{R}^d$ |
| Step 1 | Initial $k$ cluster center $C = \{c_1, c_2, \ldots, c_k\} \in \mathbb{R}^d$ is randomly picked |
| Step 2 | For each $i \in \{1, \ldots, k\}$, the cluster $C_i$ is set to be the set of points that are closer to $C_i$ than they are to $C_j$ for all $j \neq i$ |
| Step 3 | For each $i \in \{1, \ldots, k\}$, $c_i$ is set to be the center of mass of all points in $C_i$ $C_i = \frac{1}{\lvert C_i \rvert} \sum_{p \in C_i} P$ |
| Step 4 | Steps 2 and 3 are repeated until $C$ no longer changes |

The main objective of choosing $k$ centers $C$ is to minimize the potential function:

$$\phi = \sum_{p \in P} \min_{c \in C} [d(p, C)]^2 \tag{2}$$

However, finding cluster center $C$ to minimize the potential function is Non-deterministic Polynomial-time (NP) hard problem. This is due to the usual iterative method of Lloyd's algorithm in this clustering which is unlikely to get the minimum error for the cluster centers. In general, this may work but there is no theoretical guarantee of getting close to the minimum. As a solution, random values are chosen in this clustering and logically, this approach may work but without assurance of minimizing the error. Another problem with the random initialization is the risk of having two centers $c_i$ and $c_j$ in the same cluster $C_i = C_j$, that will produce overlapping parallelepipeds. The description of this issue could be seen in Fig. 2. From Fig. 2, three cluster centers $C_1$, $C_2$ and $C_3$ are randomly picked. Based on the result, we could see that the cluster centers $C_1$ and $C_2$ are actually appointed from the same cluster.

Thus, with the new edition of this algorithm the issue of random picked of cluster center initialization is solved by spreading out the centers. In this approach,

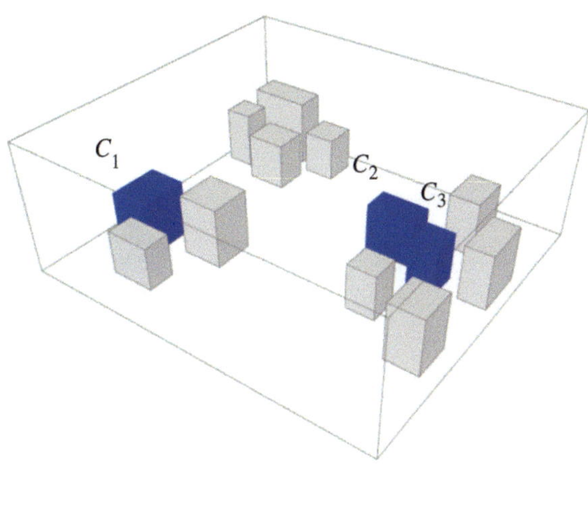


**Fig. 2** Random initialization of $k$-means crisp clustering

one center $C_1$ is chosen uniformly from $P$. Then, a new center $C_i$ is defined based on the probability of

$$\frac{D(x)^2}{\sum_{x \in X} D(x)^2} \tag{3}$$

The new edition of $k$-means algorithm is known as $k$-means++ in Arthur and Vassilvitskii (2007). $k$-means++ improved the original algorithm by defining the initial cluster center. Initial seeds are defined based on the farthest apart. Then, remaining data are clustered based on the nearest distance to the initial seeds. The initial seeding in $k$-means++ is proven to yield the improvement from its original algorithm. The algorithm of 3D $k$-means++ is described as follows.

| | |
|---|---|
| Input | a set of 3D vector data $P = \{p_1, p_2, \ldots, p_n\} \in \mathbb{R}^d$ |
| Step 1 | Choose initial center $C_1$ |
| Step 2 | Choose a new center $C_i$, by choosing $p \in P$ with probability $\frac{D(p)^2}{\sum_{p \in P} D(p)^2}$ (4) |
| Step 3 | *Step 2* is continued until $k$ centers $C_1, \ldots, C_k$ are chosen |
| Step 4 | Proceed with the standard $k$-means approach |

The same dataset as depicted in Fig. 2 has been tested using the new edition of $k$-means algorithm. From the result, the cluster centers $C_i$ are uniformly spread (see Fig. 3) and the probability of having two or more center in one cluster is reduced. Based on the results from Figs. 2 and 3, the parallelepipeds produced for each method of 3D R-Tree are described in Table 1. Based on these results, 3D $k$-means++ produced a better 3D R-Tree structure with no overlapping parallelepiped.

Based on its simplicity and lower risk of choosing center from the same cluster, we adopted 3D $k$-means++ crisp clustering to be utilized in 3D R-Tree construction

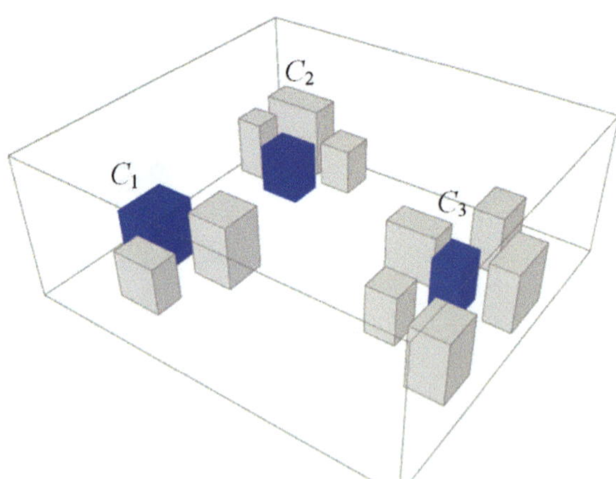


**Fig. 3** Uniformly spread cluster centers $C_i$

**Table 1** Produced structure of 3D R-tree using different approaches

| | Current Approach | Proposed 3D Crisp Clustering |
|---|---|---|
| Clustering Results | 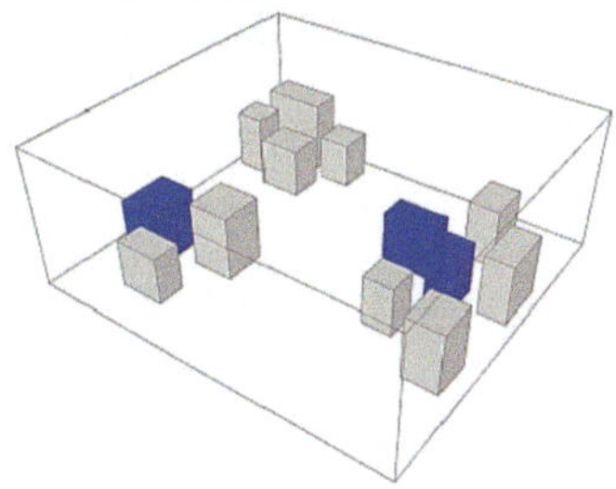 | 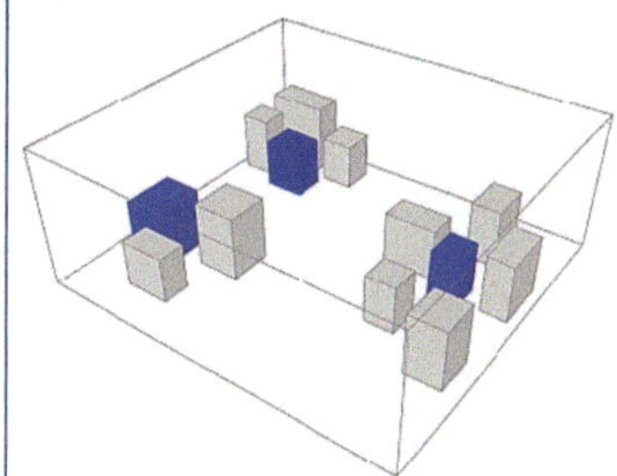 |
| 3D R-Tree | 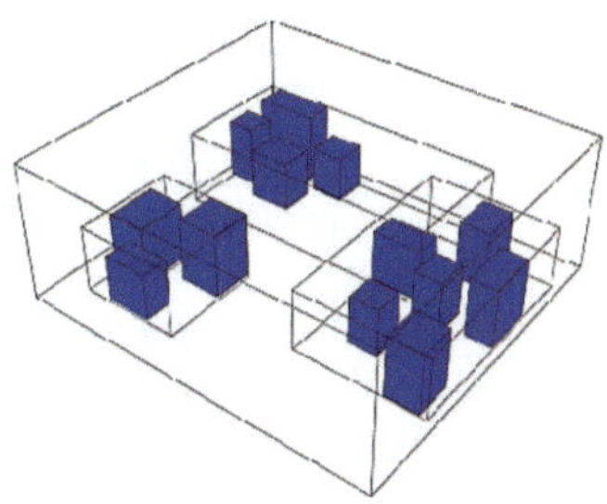<br>Number of produced node = 3<br>Number of overlapping node = 2 | 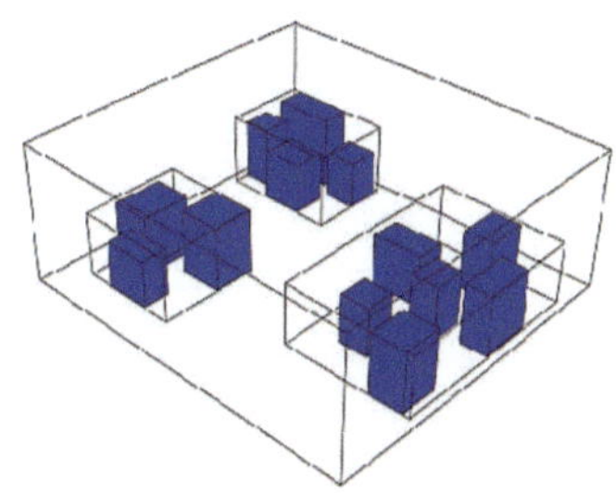<br>Number of produced node = 3<br>Number of overlapping node = 0 |

and splitting operation. The workflow of 3D R-Tree construction based on crisp clustering approaches is illustrated in Fig. 4. The test and results of this implementation are discussed in the following section.

# 4 Analysis and Results

To find the potential and efficiency of this algorithm for 3D R-Tree construction, a set of vector data needs to be tested and quantized. Datasets used in this research are a set of 3D building blocks with Level of Detail 2 (LoD 2) of CityGML definition. By following CityGML the standard, these 3D building blocks are then clustered using 3D *k*-means++ crisp clustering algorithm. The cluster classes are defined based on the maximum number of entries $M$. Clustered blocks are then grouped into parallelepipeds and each parallelepiped with the overflown node $N$ is qualified for the next cycle. In our experiment, the maximum number of 40 building blocks ($n = 40$) is demonstrated using the proposed crisp clustering algorithm. As an initial

**Fig. 4** Workflow of 3D R-Tree construction based on crisp clustering

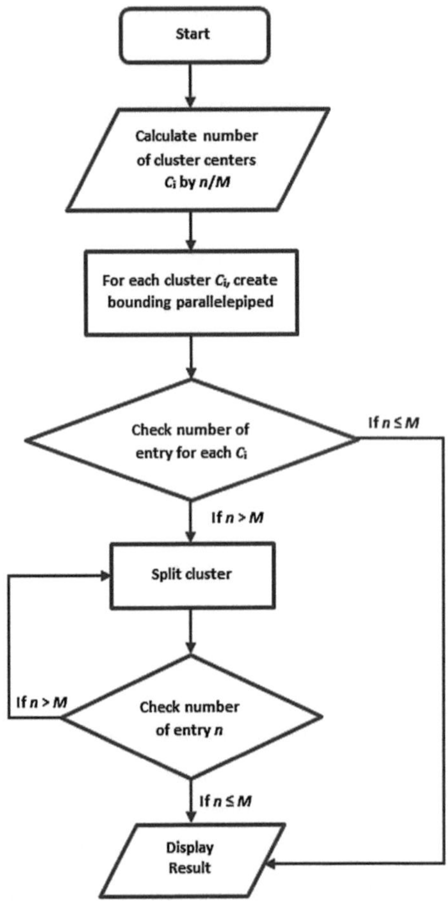


procedure, the maximum number of entries is initialized with $M = 6$. Later, $n$ building blocks are clustered based on the value of $M$, that is to say, that each clustered group could only hold at most $M$ objects. In this experiment, blocks will be grouped into six clusters.

As a result, building blocks are categorized into six groups of 3D R-Tree parallelepipeds named P, Q, R, S, T and U (see Fig. 5) based on their geographic location or position as exemplified in Fig. 5. However, among these groups there are two parallelepipeds (P and T) holding more than M entries. Thus, nodes P and T are entitled for the next cycle. In the second cycle, each node will be split again using the crisp clustering algorithm and divided into two sub-groups of P (P1 and P2) and T (T1 and T2). Overall results of this dataset are explained in Fig. 6.

Another experiment conducted in this research is to test the efficiency of the proposed crisp clustering method in reducing the number of overlapping nodes. For this experiment, a set of 3D vector data is indexed using the 3D R-Tree structure. Objects are indexed using three different approaches which are the original R-Tree,

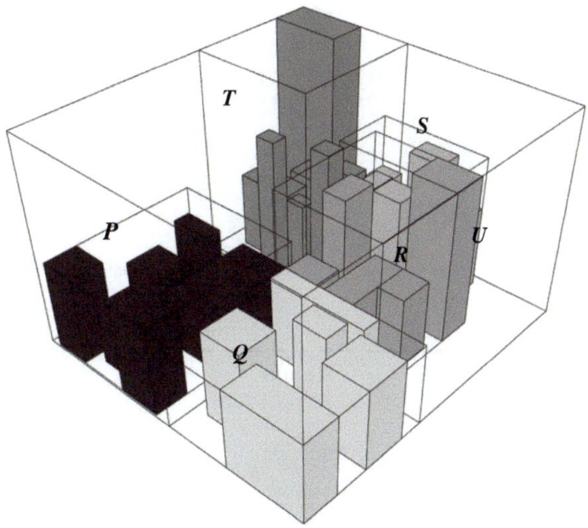


**Fig. 5** 3D R-Tree based on crisp clustering

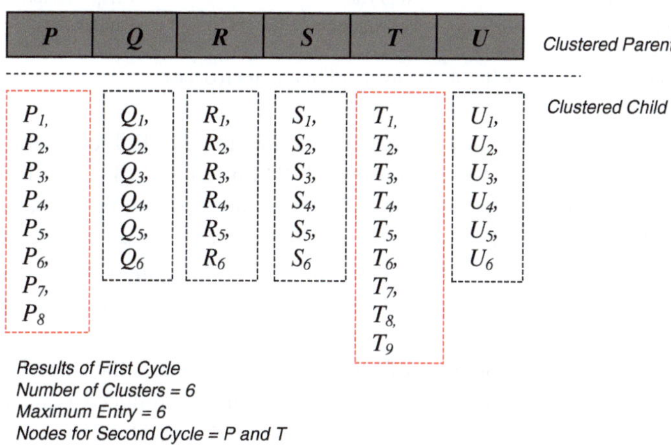


**Fig. 6** Clustered objects in 3D space

the new linear splitting by Ang/Tan and our new 3D *k*-means++ crisp clustering. Then, each result is compared based on total overlapping percentage for all nodes. Figure 7 shows the construction of 3D R-Tree using different approaches. Solid parallelepiped in Fig. 6 represents the overlapping nodes while wireframe parallelepipeds are non-overlapping nodes. From the experiment, the percentage of total overlapping volume for original R-Tree is 35 %, Ang/Tan 29 % and crisp clustering is only 3 %.

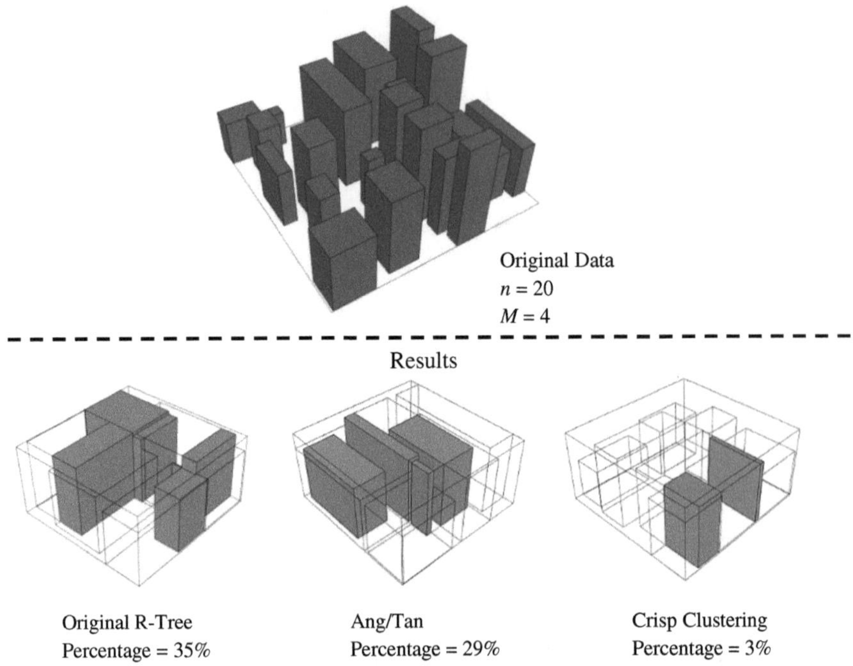


**Fig. 7** Percentage of overlapping nodes using different approaches

Another important benchmark in constructing the 3D R-Tree is the coverage volume of parallelepipeds. Several groups of 3D building blocks have been used in this experiment. Each group contain different number of building blocks which is A = 200 blocks, B = 500 blocks, C = 1,000 blocks, D = 10,000 blocks and E = 30,000 blocks. Based on the experiment, crisp clustering offers lower volume coverage than other two techniques. The result of this experiment is described in the following Fig. 8.

In order to verify the efficiency of crisp clustering towards the size of 3D R-Tree structure, the tree height and number of parallelepiped produced is measured and analyzed. A set of 3D building blocks with $n = 30{,}000$ and $M = 25$ is indexed using 3D R-Tree with a different approaches. Height of the tree is measured and total number of node is calculated. The height and number of produced node should be low in order to minimize the traversal exploration from the root node to the leaf node. Table 2 presents a comparison of height and total number of nodes using different approaches.

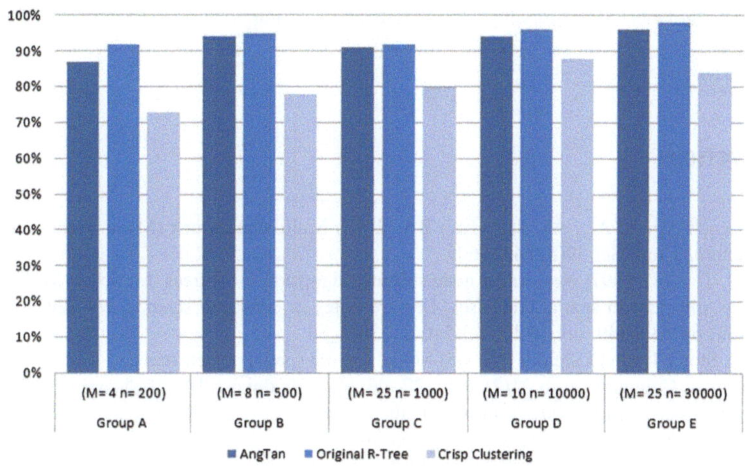


**Fig. 8** Comparison of volume coverage percentage based on several approaches

**Table 2** Produced number of nodes using different approaches

| Method | Height | Number of node |
|---|---|---|
| Ang/tan | 4 | 1,222 |
| Linear | 4 | 1,226 |
| Exhaustive | 4 | 1,230 |
| Crisp clustering | 4 | 1,200 |

## 5 Conclusions

For a massive 3D geospatial data application, retrieving information could be a crucial task and issue in data management. Practically, by implementing spatial indexing with the query operation, information will be organized sufficiently and the information will be retrieved efficiently. However, in some cases of spatial indexing structure such as 3D R-Tree, several factor need to be fine-tune according to several factors such as overlapping node, coverage area and splitting operation.

In this paper, we have proposed 3D crisp clustering algorithm for the construction of 3D R-Tree. Due to the comprehensive consideration of overlapping parallelepipeds, volume coverage and splitting operation, the crisp clustering algorithm proposed in this paper demonstrated the practically in minimizing the overlapping nodes and volume coverage. Besides that, the produced number of nodes based on the 3D crisp clustering method is smaller than other methods. Furthermore, based on the test of coverage, overlapping and produce number of nodes in this paper, our proposed crisp clustering algorithm method offers a new potential and promising approach for 3D R-Tree construction. For future research, we plan to extend the experiment and analysis by proofing the proposed algorithm for performance testing comparison towards the existing approaches.

**Acknowledgments** Major funding for this research was supported by the Ministry of Education Malaysia.

## References


Al-Badarneh AF, Al-Alaj AS, Mahafzah BA (2013) Multi small index (MSI): a spatial indexing structure. J Info Sci 39(5):643–660

Ang CH, Tan TC (1997) New linear node splitting algorithm for R-trees. In: Scholl M, Voisard A (eds) Advances in spatial databases. Lecture notes in computer science, vol 1262. Springer, Berlin, pp 337–349. doi:10.1007/3-540-63238-7_38

Arens C, Stoter J, van Oosterom P (2005) Modelling 3D spatial objects in a geo-DBMS using a 3D primitive. Comput Geosci 31(2):165–177. doi:http://dx.doi.org/10.1016/j.cageo.2004.05.013

Arthur D Vassilvitskii S (2007) k-means++: the advantages of careful seeding. In: Proceedings of the eighteenth annual ACM-SIAM symposium on discrete algorithms. Society for industrial and applied mathematics, pp 1027–1035

Balasubramanian L, Sugumaran M (2012) A state-of-art in R-tree variants for spatial indexing. Int J Comput Appl 42(20)

Fu Y, Teng J-C (2002) Subramanya S node splitting algorithms in tree-structured high-dimensional indexes for similarity search. In: Proceedings of the 2002 ACM symposium on applied computing. ACM, pp 766–770

Guttman A (1984) R-trees: a dynamic index structure for spatial searching. SIGMOD Rec 14 (2):47–57. doi:10.1145/971697.602266

Izham MY, Uznir U, Alias AR, Ayob K (2010) Georeference, rainfall-runoff modeling and 3D dynamic simulation: physical influence, integration and approaches. In: ACM, 1st international conference and exhibition on computing for geospatial research and application, Washington, DC. 1st international conference and exhibition on computing for geospatial research and application, COM.Geo 2010

Jazayeri I (2012) Trends in 3D land information collection and management. In: Rajabifard A, Williamson I, Kalantari M (eds) A national infrastructure for managing land information. University of Melbourne, pp 81–87

Kamel I, Faloutsos C (1994) Hilbert R-tree: an improved R-tree using fractals. Paper presented at the proceedings of the 20th international conference on very large data bases

Korotkov A (2012) A new double sorting-based node splitting algorithm for R-tree. Program Comput Softw 38(3):109–118

Liu Y, Fang J, Han C (2009) A new R-tree node splitting algorithm using MBR partition policy. In: 17th international conference on geoinformatics, 2009. IEEE, pp 1–6

Liu Y, Liu G, He Z (2010) Spatial index technology for multi-scale and large scale spatial data. In: 18th international conference on geoinformatics, 2010. IEEE, pp 1–4

MacQueen J (1967) Some methods for classification and analysis of multivariate observations. In: Proceedings of the fifth Berkeley symposium on mathematical statistics and probability, California, USA, p 14

Manolopoulos Y, Nanopoulos A, Papadopoulos AN, Theodoridis Y (2006) Rtrees: theory and applications. Springer, Heidelberg

Paar P (2006) Landscape visualizations: applications and requirements of 3D visualization software for environmental planning. Comput Environ Urban Syst 30(6):815–839. doi:http://dx.doi.org/10.1016/j.compenvurbsys.2005.07.002

Sleit A (2008) On using B+-tree for efficient processing for the boundary neighborhood problem. Paper presented at the WSEAS transactions on systems

Sleit A, Al-Nsour E (2014) Corner-based splitting: an improved node splitting algorithm for R-tree. J Info Sci. doi:10.1177/0165551513516709

Sugihara K, Shen Z (2012) Automatic generation of virtual 3D city models for urban planning. In: Geospatial techniques in urban planning. Advances in geographic information science. Springer, Berlin, pp 265-283. doi:10.1007/978-3-642-13559-0_13

Uznir U, Anton F, Suhaibah A, Rahman A, Mioc D (2013a) Improving 3D spatial queries search: newfangled technique of space filling curves in 3D city modeling. In: 8th 3D geoinfo conference and ISPRS WG II/2 workshop

Uznir U, François A, Alias AR (2013b) Unified data model of urban air pollution dispersion and 3D spatial city model: groundwork assessment towards sustainable urban development for Malaysia. J Environ Prot 4(7):701–712. doi:10.4236/jep.2013.47081

Wand M, Berner A, Bokeloh M, Fleck A, Hoffmann M, Jenke P, Maier B, Staneker D (2007) Schilling a interactive editing of large point clouds. In: SPBG, pp 37–45

Wang Y, Sheng Y, Zhou L, Guo F, Zhao L (2010) An underground space object-oriented three-dimensional hybrid spatial indexing method. In: 18th international conference on geoinformatics, 2010. IEEE, pp 1–5

Xu Z, Coors V (2012) Combining system dynamics model, GIS and 3D visualization in sustainability assessment of urban residential development. Build Environ 47(0):272–287. doi: http://dx.doi.org/10.1016/j.buildenv.2011.07.012

Zhu Q, Gong J, Zhang Y (2007) An efficient 3D R-tree spatial index method for virtual geographic environments. ISPRS J Photogram Remote Sens 62(3):217–224. doi:http://dx.doi.org/10.1016/j.isprsjprs.2007.05.007

Zlatanova S (2000) 3D GIS for urban development. International Institute for Aerospace Survey and Earth Sciences (ITC)

# A Hybrid Approach Integrating 3D City Models, Remotely Sensed SAR Data and Interval-Valued Fuzzy Soft Set Based Decision Making for Post Disaster Mapping of Urban Areas

**Iftikhar Ali, Aftab Ahmed Khan, Salman Qureshi, Mudassar Umar, Dagmar Haase and Ihab Hijazi**

**Abstract** This chapter introduces a novel technique for post-disaster mapping and disaster scale estimation based on an integrated framework of SAR remote sensing and a 3D city model database, optical remote sensing imagery is used for validation purpose. SAR based urban damage detection is well established and has been used for many years. We have showed that how the existing 3D City Model can be updated with the information extracted from satellite remote sensing data. The third dimension will play a very crucial role in evacuation rout planning of damaged or affected buildings. In this study in the proposed–three-level ($L_1$, $L_2$, $L_3$)–model damage assessment information is integrated with the semantic knowledge of 3D city models in order to better organize the search and rescue operation. $L_1$ includes

I. Ali (✉)
Department of Geography, University College Cork, Cork, Ireland
e-mail: iffi.math@gmail.com

A.A. Khan
Technical University Munich, Munich, Germany
e-mail: aftab.khan@tum.de

S. Qureshi · D. Haase
Department of Geography, Humboldt University of Berlin, Unter den Linden, Berlin, Germany
e-mail: salman.qureshi@geo.hu-berlin.de

D. Haase
e-mail: dagmar.haase@ufz.de

M. Umar
Department of Remote Sensing and Geoinformatics, Institute of Space Technology, 75270 Karachi, Pakistan
e-mail: mudassar_bzu@yahoo.com

I. Hijazi
An-Najah University, Nablus, Palestine
e-mail: eehab@najah.edu

© Springer International Publishing Switzerland 2015

M. Breunig et al. (eds.), *3D Geoinformation Science*,
Lecture Notes in Geoinformation and Cartography,
DOI 10.1007/978-3-319-12181-9_6

remotely sensed Synthetic Aperture Radar (SAR) space-borne data collection from the affected areas; $L_2$ includes a change detection process; and $L_3$ includes the estimation of the most affected building(s). Using this model, we show how the day-night image acquisition capability of a SAR sensor and semantic information from a 3D city model can be effectively used for post disaster mapping for rapid search and rescue operations. For $L_1$, the combination of very high resolution SAR data and a 3D city model in CityGML format is used. $L_2$ works under predefined conditions to detect the types of changes that have occurred. In $L_3$, an interval-valued fuzzy soft set theory method is proposed with which to estimate the scale of damage to the affected structures (buildings). In this study, we show the potential application of existing 3D city models (with semantic knowledge) in combination with SAR remote sensing for post disaster management activities, especially for search and rescue operations.

**Keywords** Urban remote sensing · Change detection · CityGML · Hazard mapping · Spatial analysis

## 1 Introduction

Urban areas with high concentration of people and multi-story residential apartments may result in high numbers of causalities in a building collapse following an earthquake. Earthquakes are highly unpredictable events, so that early warnings are required for successful evacuation (Kagan 1997). Earthquakes belong to the most deadly and destructive class of disasters where the delay in response could cause increase in number of causalities. With the rate of urbanization increasing as a result of population (and/or economic) growth and industrialization, global warming is also increasing. In future, for natural disasters management, more efficient, scientifically reliable and rapid search and rescue techniques will be required.

During disasters (earthquakes) in highly populated urban areas, the response time for emergency services is a crucial factor and should be optimized, while at the same time, the process of identification of damaged areas needs to be more fast, precise and as early as possible (Trianni and Gamba 2009). In the past, traditional methods such as aerial imagery and visual image interpretation (Lu et al. 1998) have been used for damage detection. For example, Suga et al. (2001) used InSAR data to detect urban damages and land displacement. Since the early 1990s, airborne laser scanning has been actively used for the change detection (Martin and Bernhard 2010; Murakami et al. 1999; Vogtle and Steinle 2004). In earthquakes in highly populated cities, the rapid estimation of potential damage and the detection of affected areas on a large scale have been a great challenge. Despite their advantages, the techniques that have traditionally been used have several limitations. More recently cities (and, more specifically, buildings) are now being analyzed in 3D virtual reality environment using semantic information from these structures. The use of semantic 3D building models with international open standards such as CityGML and Industry Foundation Classes (IFC) has become very common. In recent years, many cities throughout the world have been switched to these international standards to support semantics and efficient visualization and to fulfill the challenging demands of users. CityGML is a common semantic information model for the representation, storage, and exchange of 3D urban objects based on the XML format (Kolbe et al. 2005). CityGML is implemented as an application schema of Geography Markup Language version 3.1.1 (GML3) (Kolbe et al. 2005) and is an international standard for the representation and exchange of semantic 3D city models (CityGML 2014; Gröger et al. 2008; Kolbe 2009). CityGML-based 3D city models work as the central information hub for various domains, such as tourism, environmental projection, urban planning, and navigation (Kolbe 2009). From a disaster management perspective, 3D city models in CityGML contain rich information, representing semantics, geometry, topology, and appearance (Kolbe et al. 2005), and objects can be represented at up to five levels of detail (LOD) (Kolbe 2009). Kolbe et al. (2005, 2009) gives a comprehensive overview of 3D city modeling and CityGML.

Remote sensing is another way of gathering information at a large scale. Recent development in earth observation sensors has made it possible to acquire high-resolution data at large scale. COSMO-SkyMed (Italian) and TerraSAR-X (German) are very high-resolution X-band Earth observation systems that are equipped with multi-acquisition modes, and they have large wide swath image acquisition capability. COSMO-SkyMed has a revisit time of $\approx 4.5$ h with full constellation of four satellites, whereas TerraSAR-X has an 11-day revisit time. In the future, such sensors could be used for near real time monitoring activities. In the field of remote sensing, the major constraint is bad weather conditions; in such cases, SAR is the only reliable source of acquiring wide-swath and multi-temporal imagery. Number of studies has been done on the utilization of SAR data for the detection of change/damages caused by earthquake (Brunner et al. 2010; Dell'Acqua and Gamba 2012; Dell'Acqua et al. 2010, 2009; Dong and Shan 2013; Park et al. 2013; Tong et al. 2013; Trianni and Gamba 2009).

To the best of our knowledge, there is no framework that exploits the existing *3D Semantic City Model* and the SAR remotely sensed data for disaster management and mapping. The objective of this paper was to establish the feasibility of the mechanism that combines remotely sensed SAR data and a 3D Semantic City Model database, so that, in any disaster, the database for each building in a 3D Semantic City Model can be updated with additional information, and then query based semantic filtering (Kolbe et al. 2005) of damaged buildings can be used. Soft sets, fuzzy sets and fuzzy soft sets are the recent methods that have been developed and have matured to address the uncertainties present in real world situations (Chetia and Das 2010). In this paper, we propose the application of *interval-valued fuzzy soft sets* for the estimation of the scale of damage; a detailed description is provided in Sect. 5.3.

**Table 1** Data used for this study

| Sensors | Date | Event | Specifications |
|---|---|---|---|
| TerraSAR-X | 05-12-2007 | Pre-event ($X_{t_1}$) | Mode: SpotLight, polarization: HH, incidence angle: 39.8° |
| TerraSAR-X | 02-10-2011 | Post-event ($X_{t_2}$) | Mode: SpotLight, polarization: HH, incidence angle: 39.8° |
| QuickBird | 11-11-2007 | Pre-event ($X_{t_1}$) | Mode: PAN |
| QuickBird | 09-10-2011 | Post-event ($X_{t_2}$) | Mode: PAN |

## 2 Data Used, Study Area and Preprocessing

The area under investigation is San Francisco, USA. The TerraSAR-X image pair was received at the L1B product level with a pixel spacing of 0.45 and 0.87 m in the slant range and azimuth, respectively. One image from each of QuickBird and WorldView-2 (PAN format) was used for cross check/validation. Table 1 shows the details of the data used.

To obtain square ground pixels, the SAR images were multi-looked. The images were co-registered, and a subset was taken to expedite the processing chain. Following the multi-looking and co-registration, an enhanced lee filter was applied to reduce the speckle effect. Geo-coding and radiometric correction were applied, and the average value of the backscattering coefficient was calculated (on a logarithmic scale) by overlaying the digital shape files of the building footprint. The image log ratio $[X_{LR} = \log(X_{t_2}/X_{t_1})]$ technique was applied for the identification of changes. Other change detection methods can be used for change detection analysis (Lu et al. 2004). The scheme for the preprocessing steps of the TerraSAR-X images is illustrated in Fig. 1.

## 3 Methodology

### 3.1 Proposed Model

The LoD3 (level of detail) is sufficient for the proposed methodology where the building's architectural information like height, use (commercial or residential) number of stories are given; which are very crucial for search and rescue operations. We divided our model theoretically into three levels of implementation, which are

$L_1$ Space-borne data collection—information gathering regarding the affected areas

$L_2$ Change detection types (CDT)—identification of the affected area

$L_3$ Prediction—estimation of possible damage type and scale based on updated database

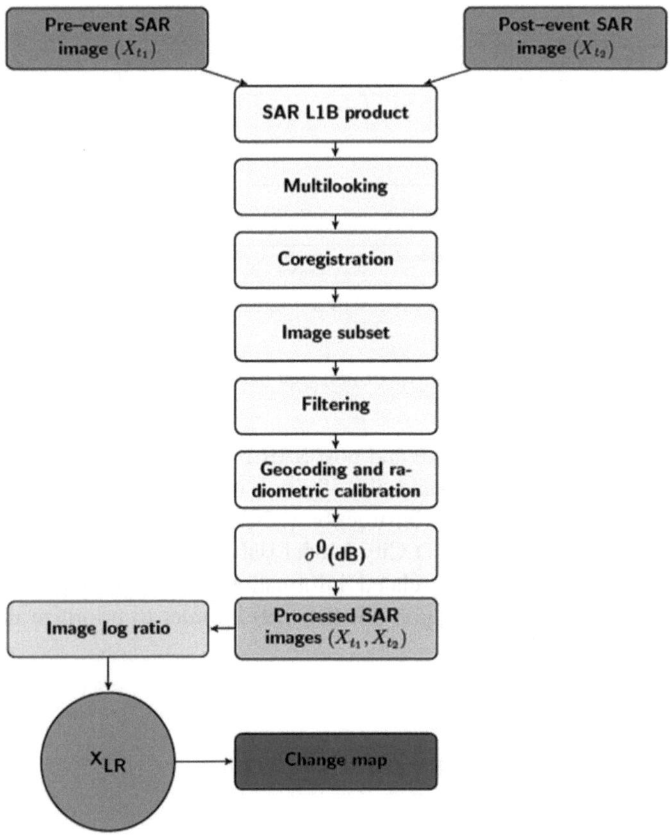


**Fig. 1** Flowchart for SAR data preprocessing

A detailed description of each of these levels is given below (Sects. 3.2, 3.3 and 3.4).

### 3.1.1 Workflow of Proposed Model

The proposed model (Fig. 2) will start from Level $L_1$ where the labels (numbering in red color in Fig. 2) **1** (remote sensing data) and **4** (3D City Model database) represent the input data required for this model. After preprocessing the remote sensing data (label **2**) building footprints will be converted to ROIs (label **5–6**) in order to calculate the average signal backscatter for these ROIs (label **3**) and 3D City Model database (label **4**) will be updated with the pre and post event signal backscatter information for each building (this step does not mean that from now to onward model is fully shifting from 3D to 2D; rather this is just the one information extracted from 3D City Model in order to further update the existing 3D City

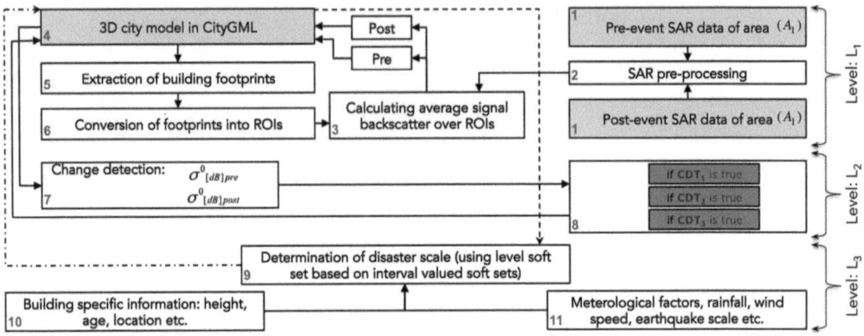


**Fig. 2** $L_1$, $L_2$ and $L_3$ are the three levels of proposed model

Model). In the next step both pre and post SAR backscatter will be further use for the change detection (label **7**) and classification of buildings in the database into destroyed, no change and new construction (label **8**). In the final step information about destroyed (affected) in 3D City Model (label **4**) along with meteorological (label **11**) and other building related information (label **10**) would be used in *interval-valued fuzzy soft set* algorithm (label **9**) in order to prioritize the damage scale of each building.

## 3.2 Level $L_1$: Space-Borne Data Collection

The basic data requirements for L1 and this proposed model are a 3D city model and spaceborne VHR-SAR imagery. First, the footprints of all buildings (or the most important residential skyscrapers) were selected, and then these footprints were converted to regions of interest (ROIs). The pre-event SAR image acquisition at time $t_1$ was used in combination with the ROIs to calculate the average back scattering value ($\sigma^0$) over each ROI (representing building footprints). An attribute table for each building, containing semantic and other building related information, was edited to include two additional attributes: the pre-event ($\sigma^0_{[dB]pre}$) value and the post-event ($\sigma^0_{[dB]post}$) value. If an event (earthquake) happened at time $t_{1,}$ then the SAR image acquired after the event is represented as $t_2$. Using the same ROIs, which were used to calculate pre-event ($\sigma^0_{[dB]pre}$) values, the post-event ($\sigma^0_{[dB]post}$) values were calculated and were stored in the edited attribute table of the 3D building model. In Fig. 2, the workflow mechanism of $L_1$ is shown.

## 3.3 Level $L_2$: Change Detection Types (CDT)

At this level, the 3D city model (along with the semantic, geometric and topological information) has the average backscattering value of the pre-event ($\sigma^0_{[dB]pre}$) and post-event ($\sigma^0_{[dB]post}$) for each building (or selected buildings) in the 3D city model database table. The input required for $L_2$ is ($\sigma^0_{[dB]pre}$) and post-event ($\sigma^0_{[dB]post}$). At $L_2$, detailed analysis is performed in order to categorize the change detection types (CDT) into three major types, which are:

I. $CDT_1$: Destroyed—the buildings were considered to be affected by the disaster if

$$\sigma^0_{[dB]pre} > \sigma^0_{[dB]post}$$

The level $L_3$ will be executed if and only if $CDT_1$ is true.

II. $CDT_2$: No change—the building (or area) is safe from the disaster if

$$\sigma^0_{[dB]pre} \approx \sigma^0_{[dB]post}$$

is true.

III. $CDT_3$: New construction—if there is an empty plot next to a building $b_1 \in \{the\,existing\,3D\,citymodel\}$, one can confirm that this area (plot) is still empty or whether if there is a new building in that plot. If the condition

$$(\sigma^0_{[dB]pre} - \sigma^0_{[dB]post}) > -5[dB]$$

is true, that is an indicator for urbanization or new construction. The average radar signal backscattering from man-made structures (urban) is greater than −5(dB) (Dekker 2010), so this finding can be integrated to detect urbanization. Figure 2 describes the workflow of $L_2$.

## 3.4 Level $L_3$: Prediction

Following the identification of the affected areas ($CDT_1$), we used an interval-valued fuzzy soft set theory based algorithm proposed by Feng et al. (2010) with some modification to predict the most affected building. Figure 2 illustrates the workflow of $L_3$. In Sect. 4, we showed the potential application of *interval-valued fuzzy soft sets* for the prediction of the scale of damage. In Sect. 5.3.2, an example application (based on hypothetical data) of this level is shown.

## 4 Implementation Scheme

Before implementing this conceptual model, we create two generic attributes, $\sigma^0_{[\mathrm{dB}]\mathrm{pre}}$ and $\sigma^0_{[\mathrm{dB}]\mathrm{post}}$, in the building table of the 3D city model database. Here, we assumed that we already have a $\sigma^0_{[\mathrm{dB}]pre}$ value taken at time ($t_1$) for the buildings under investigation. Following the event (earthquake), this (Fig. 3) workflow begins. First, we take the ROIs (building footprints) from the 3D city database and export them to the CityGML file through the CityGML export/import tool. The CityGML file is converted to an ESRI shape file using the FME tool, and the footprints are imported using the conversion function in ArcCatalog (ArcGIS). The footprints of the buildings are integrated (manually checked and parallel layered with the bounding boxes of the same buildings) and then are used to calculate the $\sigma^0_{[dB]post}$ value at acquisition ($t_2$) for each building. After calculating the $\sigma^0_{[\mathrm{dB}]post}$ value for each building, an Excel sheet is generated containing the building ID and

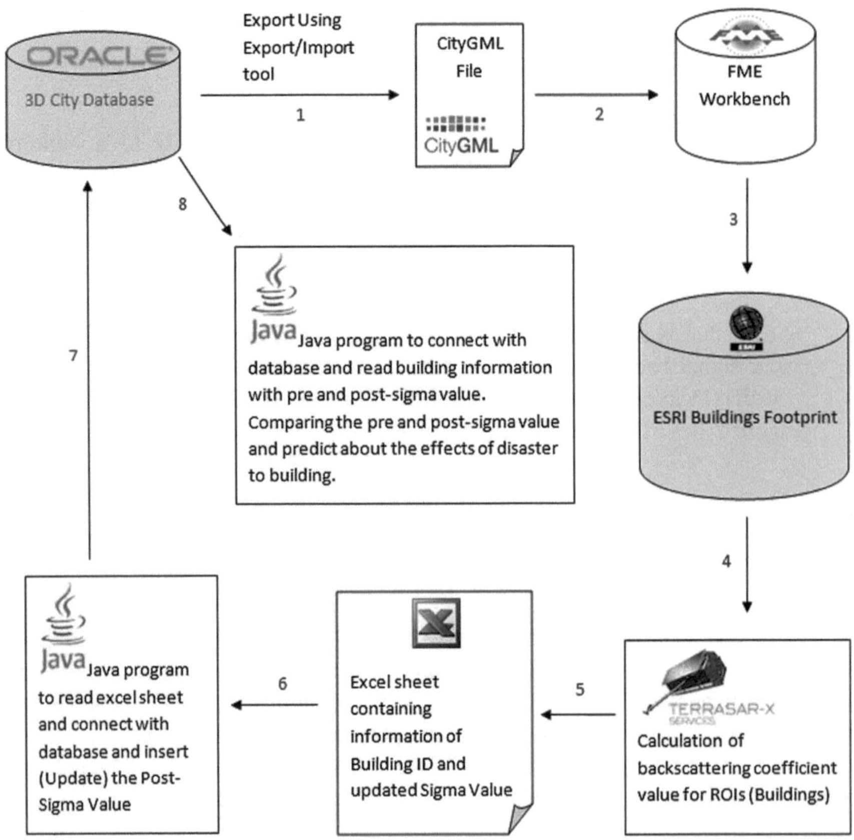


**Fig. 3** Implementation scheme

the updated $\sigma^0_{[dB]post}$ value. Through a Java program, the Excel sheet is read and connected with the 3D city database, and the $\sigma^0_{[dB]post}$ value for the relevant building is updated. Once the $\sigma^0_{[dB]pre}$ and $\sigma^0_{[dB]post}$ values for the buildings of concern are acquired, the program reads both values and classifies the buildings according to the CDT as a final step. The detailed implementation scheme is shown in Fig. 3.

# 5 Results

## 5.1 Level $L_1$: Space-Borne Data Collection

At this initial level, space-borne pre-event ($X_{t_1}$) and post-event ($X_{t_2}$) SAR data were collected and processed as described in Sect. 2. For testing the proposed model, we selected 13 ROIs in the region, and the backscattering coefficients ($\sigma^0$) were calculated for both the $t_1$ and $t_2$ data sets, which were used as the input ($\sigma^0_{[dB]pre}$, $\sigma^0_{[dB]post}$) into $L_2$.

## 5.2 Level $L_2$: Change Detection Types (CDT)

At $L_2$, the input ($\sigma^0_{[dB]pre}$, $\sigma^0_{[dB]post}$) values were analyzed in order to classify them into their corresponding CDT categories. Table 2 shows the list of $CDT_i$ types and the number of cases found for each CDT.

In Fig. 4a, three cases (row wise, $D_1$, $D_2$, $D_3$) of $CDT_1$ are shown in which structures (buildings) were present in the first acquisition ($t_1$) and disappeared in the second acquisition ($t_2$). Figure 5a shows the significant decrease in the $\sigma^0_{[dB]post}$ value with respect to the $\sigma^0_{[dB]pre}$ value for these buildings. The $\sigma^0_{[dB]post}$ values for $D_1$, $D_2$ and $D_3$ decrease by +4.29, +6.58 and +10.32 (dB) respectively in ($t_2$). $CDT_2$ is dedicated to the identification of safe buildings or areas that did not change between the $X_{t_1}$ and $X_{t_2}$ acquisitions. Figure 4b shows the five buildings that were found to be safe out of the 13 areas investigated. Figure 5b shows that the $\sigma^0_{[dB]pre}$ and $\sigma^0_{[dB]post}$ values are approximately the same for these five buildings. Samples of buildings in order to detect the new constructions, $CDT_3$ is proposed, where the

**Table 2** Categorization of CDT

| Classification | $CDT_i$ | No. of cases (K) | Abbreviation |
|---|---|---|---|
| Destroyed | $CDT_1$ | 3 | $D_i$, $i = 1,\ldots,K$ (where, $K = 3$) |
| No change | $CDT_2$ | 5 | $NC_i$, $i = 1,\ldots,K$ (where, $K = 5$) |
| New construction | $CDT_3$ | 5 | $N_i$, $i = 1,\ldots,K$ (where, $K = 5$) |

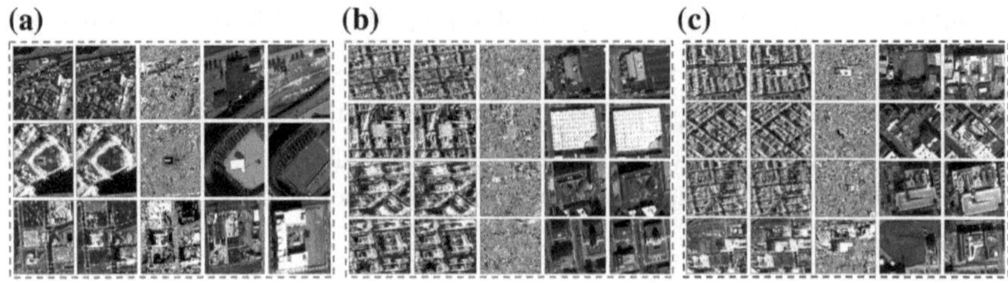


**Fig. 4** Columns 1–5 are $X_{t_1}$, $X_{t_2}$, $X_{LR}$, QuickBird (PAN) ($t_1$) and WorldView-2 (PAN) ($t_2$) respectively. **a** Rows 1–3 are the three cases which were categorized as $CDT_1$; (**b**) Column 1–5 are $X_{t_1}$, $X_{t_2}$, $X_{LR}$, QuickBird (PAN) ($t_1$) and WorldView-2 (PAN) ($t_2$) respectively. Rows 1–4 is the cases that were categorized as $CDT_2$; (**c**) Column 1–5 are $X_{t_1}$, $X_{t_2}$, $X_{LR}$, QuickBird (PAN) ($t_1$) and WorldView-2 (PAN) ($t_2$) respectively. Rows 1–4 are the four cases that were categorized as $CDT_3$

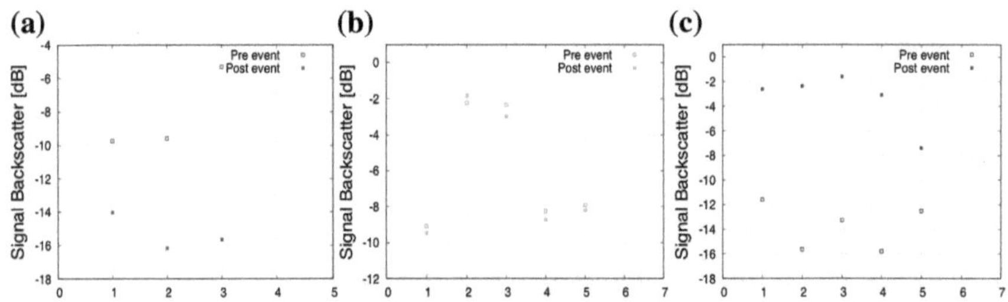


**Fig. 5** The mean backscattering value on building footprints, (**a**) $CDT_1$: Backscattering coefficient value for three ($D_1$, $D_2$ and $D_3$) cases as shown in Fig. 4a. (**b**) $CDT_2$. : Backscattering coefficient value for five cases as shown in Fig. 4b. (**c**) $CDT_3$.: Backscattering coefficient valufor five cases as shown in Fig. 4c

value of $\sigma^0_{[dB]post}$ must be greater than that of $\sigma^0_{[dB]pre}$. Five newly constructed buildings were found in the region, as shown in Fig. 4c. The value of $\sigma^0_{[dB]post}$ for safe buildings (NC) increases by 8:97, 13:27, 11:66, 12:70 and 11:79 (dB) compared with the $\sigma^0_{[dB]pre}$, as shown in Fig. 5c.

### 5.3 *Level L3: Prediction*

At $L_3$, we proposed the application of interval-valued fuzzy soft sets for the prediction of the scale of damage. This level will be executed if and only if $CDT_1$ is true for $L_2$. In this section, we show the application framework for this procedure with an example demonstration.

### 5.3.1 Preliminaries: Interval-Valued Fuzzy Soft Sets

Let $U$ be the initial universe and $E$ be the set of parameters. Let $P(U)$ denote the set of all subsets and $A \subseteq E$.

**Definition 1** (Maji et al. 2002, Molodtsov 1999) A pair $(F, E)$ is called a *soft set over U* if and only if $F$ is a mapping of $E$ into the set of subsets of the set $U$ (power set $P(U)$), i.e., $F: E \rightarrow P(U)$.

**Definition 2** (Maji et al. 2001) A pair $(G, A)$ is called a *fuzzy soft set over U* where $G$ is a mapping of $A \subseteq E$ into the set of all fuzzy subsets in $U$ denoted by $F(U)$), i.e., $G: A \rightarrow F(U)$.

**Definition 3** (Yang et al. 2009) An *interval-valued fuzzy set I* on an universe $(U)$ is a mapping such that

$$I{:}U \rightarrow Int([0,1])$$

where $Int([0,1])$ stands for the set of all closed subintervals of ([0,1]), the set of all *interval-valued fuzzy sets* is denoted by $P(U)$.

**Definition 4** (Yang et al. 2009) Let $U$ be an initial universe, $E$ be a set of parameters and $A \subseteq E$., a pair $(\dot{F}, A)$ is called an *interval-valued fuzzy soft set* (IVFS) over $U$, where $\dot{F}$ is a mapping defined as $F : E \rightarrow P(U)$.

**Definition 5** (Feng et al. 2010) Let $(U, E)$ be a st universe and $A \subseteq E$. Let $(\dot{F}, A)$ be an (IVFS) over $U$ such that $\forall$ a $\in$ A, $\dot{F}$ *(a)* is an *interval-valued fuzzy set* with $\dot{F}(a)(\mathrm{x}) = [\dot{F}(a)_*(\mathrm{x}), \dot{F}(a)^*(x)]$, $\forall \mathrm{x} \in \mathrm{U}$, where $\dot{F}(a)_*(x)$ and $\dot{F}(a)^*(x)$ are called the lower and upper degrees of membership of x in $\dot{F}$. Then the fuzzy soft set $(\dot{F}, A)$ over $U$ given by

$$\dot{F}_{-}(a) = \left\{ \left(x, \dot{F}(a)_*(x)\right) : \mathrm{x} \in U \right\}, \quad \forall a \in A$$

is called *pessimistic reduct fuzzy soft set* (PRFS) of $(\dot{F}, A)$.

Similarly,

$$\dot{F}_{+}(a) = \left\{ \left(\mathrm{x}, \dot{F}(a)^*(x)\right) : x \in \mathrm{U} \right\}, \quad \forall \mathrm{a} \in \mathrm{A}$$

is called *optimistic reduct fuzzy soft set* (ORFS) of $(\dot{F}, A)$ and

$$\dot{F}_{\mathrm{N}}(a) = \left\{ \left(x, \left(\dot{F}(a)_*(x) + \dot{F}(a)^*(x)\right)/2\right) : x \in U \right\}, \quad \forall \mathrm{a} \in A$$

is called *neutral reduct fuzzy soft set* (NRFS) of $(\dot{F}, A)$.

**Definition 6** (Feng et al. 2010) Let (*U, E*) be soft universe and $A \subseteq E$ and t ∈ [0,1]. The t-*level soft set of a fuzzy soft set* ($\dot{F}$, *A*) over *U* is at ($F_t$ (a)) defined by

$$(F_t(a)) = \{x \in U: \dot{F}(a)(x) \geq t\}, \quad \forall a \in A$$

### 5.3.2 Application Example

Suppose that, following $L_2$, we have detected that condition *C*is true for the following five buildings, $B_i \in \{3D\ CityModelling\}$ (where, *i* = 1, 2, …, 5); $CDT_1$ being true indicates that buildings are affected (or most affected) by the event. We can then predict the most severely affected building among these five buildings and then can be reorded ading to the level of destruction by using level soft sets based on *interval-valued fuzzy soft sets.*

Suppose the following:

i. $U = \{d_1, d_2, d_3, d_4, d_5\}$ = {Five affected buildings}
ii. $E = \{\varepsilon_1, \varepsilon_2, \varepsilon_3, \varepsilon_4, \varepsilon_5\}$ = {magnitude of earthquake, change in $\sigma^0$, age of building, earthquake duration, building original height}. Different lt of parameters can also be used according to the present scenario and expert opinion.

**Algorithm** (Feng et al. 2010; Yang et al. 2009).

1. *Input*: The set of *Interval-valued fuzzy soft sets* defined by using definition 4. In our case *Interval-valued fuzzy soft set* ($\dot{F}$, *A*) is shown in Table 3.
2. *Input*: Parameter set *E* consists of preferences of decision maker, parameter under consideration in our case are $E = \{\varepsilon_1, \varepsilon_2, \varepsilon_3, \varepsilon_4, \varepsilon_5\}$ = {magnitude of earthquake, change in $\sigma^0$, age of building, earthquake duration, building original height}

**Table 3** Interval-valued fuzzy soft set ($\dot{F}$, *A*)

| *U* | $\varepsilon_1$ | $\varepsilon_2$ | $\varepsilon_3$ | $\varepsilon_4$ | $\varepsilon_5$ |
|---|---|---|---|---|---|
| $d_1$ | [0.7, 1.0] | [0.5, 0.7] | [0.2, 0.5] | [0.5, 0.6] | [0.7, 0.8] |
| $d_2$ | [0.6, 0.2] | [0.9, 0.1] | [0.8, 0.9] | [0.5, 0.7] | [0.1, 0.3] |
| $d_3$ | [1.0, 0.9] | [0.3, 0.6] | [0.6, 0.5] | [0.7, 0.3] | [0.2, 0.9] |
| $d_4$ | [0.2, 0.8] | [0.7, 0.3] | [0.3, 1.0] | [0.9, 0.1] | [0.4, 0.8] |
| $d_5$ | [0.4, 0.2] | [0.4, 0.5] | [0.5, 0.8] | [0.6, 0.1] | [0.9, 0.3] |

**Table 4** Pessimistic reduct fuzzy soft set (PRFS) of ($\dot{F}$, A)

| $U$ | $\varepsilon_1$ | $\varepsilon_2$ | $\varepsilon_3$ | $\varepsilon_4$ | $\varepsilon_5$ |
|---|---|---|---|---|---|
| $d_1$ | 0.7 | 0.5 | 0.2 | 0.5 | 0.7 |
| $d_2$ | 0.2 | 0.1 | 0.8 | 0.5 | 0.1 |
| $d_3$ | 0.9 | 0.3 | 0.5 | 0.3 | 0.2 |
| $d_4$ | 0.2 | 0.3 | 0.3 | 0.1 | 0.4 |
| $d_5$ | 0.2 | 0.4 | 0.5 | 0.1 | 0.3 |

**Table 5** Optimistic reduct fuzzy soft set (ORFS) of ($\dot{F}$, A)

| $U$ | $\varepsilon_1$ | $\varepsilon_2$ | $\varepsilon_3$ | $\varepsilon_4$ | $\varepsilon_5$ |
|---|---|---|---|---|---|
| $d_1$ | 1.0 | 0.7 | 0.5 | 0.6 | 0.8 |
| $d_2$ | 0.6 | 0.9 | 0.9 | 0.7 | 0.3 |
| $d_3$ | 1.0 | 0.6 | 0.6 | 0.7 | 0.9 |
| $d_4$ | 0.8 | 0.7 | 1.0 | 0.9 | 0.8 |
| $d_5$ | 0.4 | 0.5 | 0.8 | 0.6 | 0.9 |

**Table 6** Neutral reduct fuzzy soft set (NRFS) of ($\dot{F}$, A)

| $U$ | $\varepsilon_1$ | $\varepsilon_2$ | $\varepsilon_3$ | $\varepsilon_4$ | $\varepsilon_5$ |
|---|---|---|---|---|---|
| $d_1$ | 0.85 | 0.60 | 0.35 | 0.55 | 0.75 |
| $d_2$ | 0.40 | 0.50 | 0.85 | 0.60 | 0.20 |
| $d_3$ | 0.95 | 0.45 | 0.55 | 0.50 | 0.55 |
| $d_4$ | 0.50 | 0.50 | 0.65 | 0.50 | 0.60 |
| $d_5$ | 0.30 | 0.45 | 0.65 | 0.35 | 0.60 |

3. *Compute*: By using definition 5 calculate $\dot{F}_{-}(a)$, $\forall a \in A$ (PRFS), $\dot{F}_{+}(a)$, $\forall a \in A$ (ORFS) and $\dot{F}_{N}(a)$, $\forall a \in A$ (NRFS) which are shown in Table 4, Table 5 and Table 6 respectively.
4. *Compute*: $\forall d_i \in U$, compute the interval fuzzy choice value $c_i$ for each damage type di such that

$$c_i = \left[c_i^-, c_i^+\right] = \left[\sum_{a \in A \subseteq E} \dot{F}_{-}(a)(d_i), \sum_{a \in A \subseteq E} \dot{F}_{+}(a)(d_i)\right]$$

Table 7 shows the choice values for each $d_i$.

5. *Compute*: $\forall\ d_i \in U$, compute the score value $r_i$ for each $d_i$ such that

$$r_i = \sum_{d_i \in U}\left(\left(c_i^- - c_j^-\right) + \left(c_i^+ - c_j^+\right)\right)$$

**Table 7** Choice values and scores

| $U$ | $c_i$ | $r_i$ |
|---|---|---|
| $d_1$ | $c_1$ = [2.6, 3.6] | $r_1$ = 3.3 |
| $d_2$ | $c_2$ = [1.7, 3.4] | $r_2$ = −2.0 |
| $d_3$ | $c_3$ = [2.2, 3, 8] | $r_3$ = 2.9 |
| $d_4$ | $c_4$ = [1.3, 4.2] | $r_4$ = 0.0 |
| $d_5$ | $c_5$ = [1.5, 3.2] | $r_5$ = −4.0 |

Table 7 shows the score values for each $d_i$.

6. <u>*Decision*</u>: Selection of $d_m \in U$ with maximum score value i.e., $r_m = max_{d_i \in U\{r_i\}}$. As shown in

   Table 7 $r_1 > r_i$, where i = 2, 3, 4, 5. Finally $\{d_1\}$ has the maximum score.
7. <u>*Threshold value t*</u>: We define the threshold values t on membership degree as follow

$$t = \frac{\sum_{j=1}^{N}\left(\frac{\sum_{i=1}^{N} a_{ij}}{N}\right)}{N}$$

where, t ∈ [0,1] and from Table 5 we get t = 0.55.

8. <u>*Compute*</u>: Computation of level soft set with t = 0.55 by applying definition 6 in Table 6, results are shown in Table 8.
9. <u>*Decision*</u>: By considering the aggregation *G* = *max*, from Table 4 threshold *fuzzy set* is induced which is

$$\widetilde{g} = \{\varepsilon_1 = 0.9, \varepsilon_2 = 0.5, \varepsilon_3 = 0.8, \varepsilon_4 = 0.5, \varepsilon_5 = 0.7\}$$

By using these fuzzy thresholds top-level crisp soft set is obtain as shown in Table 9. $c_1 > c_i$, where $i$ = 2, 3, 4, 5. Finally $d_1$ has the maximum choice value.

**Table 8** Level soft set with t = 0.55

| $U$ | $\varepsilon_1$ | $\varepsilon_2$ | $\varepsilon_3$ | $\varepsilon_4$ | $\varepsilon_5$ |
|---|---|---|---|---|---|
| $d_1$ | 1 | 1 | 0 | 1 | 1 |
| $d_2$ | 0 | 0 | 1 | 1 | 0 |
| $d_3$ | 1 | 0 | 1 | 0 | 1 |
| $d_4$ | 0 | 0 | 1 | 0 | 1 |
| $d_5$ | 0 | 0 | 1 | 0 | 1 |

**Table 9** Level soft set (max)

| $U$ | $\varepsilon_1$ | $\varepsilon_2$ | $\varepsilon_3$ | $\varepsilon_4$ | $\varepsilon_5$ | $c_i$ |
|---|---|---|---|---|---|---|
| $d_1$ | 0 | 1 | 0 | 1 | 1 | $c_1 = 3.0$ |
| $d_2$ | 0 | 0 | 1 | 1 | 0 | $c_2 = 2.0$ |
| $d_3$ | 1 | 0 | 0 | 0 | 0 | $c_3 = 1.0$ |
| $d_4$ | 0 | 0 | 0 | 0 | 0 | $c_4 = 0.0$ |
| $d_5$ | 0 | 0 | 0 | 0 | 0 | $c_5 = 0.0$ |

## 6 Discussion

Considering the possible future disaster threats to highly vulnerable cities, Kubal et al. (2009) used a multi-criterion approach for urban flood risk assessment. The proposed model is highly feasible for implementation and for operational search and rescue activities. The weather independence, day-night imaging capability and short revisit time of current (and future) SAR-sensors demonstrate the feasibility of implementation of this model. The proposed model has great potential to extend the database and to record the earthquake resistant history for each building. Several authors have proposed different change detection techniques that use only optical, SAR or laser scanning data. One might question the purpose of using SAR data in combination with 3D city models. For example, if an earthquake occurs on Sunday morning, then the affected areas (buildings) can be found by using SAR data; however, the important question is how to determine the priorities for rescue teams. During weekends, university, school, college, and office buildings are closed; thus, by using the semantic information query from the 3D city model, such buildings can easily be filtered out, and the focus and resources can be shifted towards the residential and other affected areas.

The advantage of using a 3D city model in combination with SAR data is that, once an affected building is identified, the rescue teams can be assisted accordingly by using the geometric and topological consistency (Gröger and Plümer 2012) information from the 3D model. For example, by analyzing the geometry and topological properties of the building (3D model), one can guide the rescue teams in real time with respect to the types of cutters, jacks or hydraulic cranes (small or large) that might be needed to rescue trapped people. Another major advantage of the combined use of an existing 3D city model database and SAR remote sensing is that the SAR change detection information can be further filtered and classified based on the 3D city model semantic information. For example, in our case, we have three buildings ($D_1$, $D_2$ and $D_3$) that were found to be destroyed as a result of the SAR change detection ($CDT_2$); by using the 3D city model semantic information, we can further query whether each building belongs to such categories as public schools, hotels, residential, or trade/office buildings. Thus, in large-scale disasters, limited resources can be used more efficiently, and search and rescue operations can be performed in a more targeted and focused manner.

Kolbe et al. (2005) (sec. 4: Application of CityGML for Disaster Management) showed that thematically rich attributes effectively allow specific queries during emergencies, allowing queries such as *"Where are buildings with flat roofs that are large enough for a helicopter to land?"* (Kolbe et al. 2005). However after an earthquake of magnitude of 8 or 9, how do we know that these buildings with flat roofs are still safe and did not collapse?" the answer will come from L2 of our proposed model and will allow more efficient relief and rescue operations. This query can now be rephrased for an emergency as, *Where are buildings with flat roofs with* $\sigma^0_{[dB]pre} \approx \sigma^0_{[dB]post}$ *(safe) that are large enough for a helicopter to land?*

The application scale of this framework could be on a local (district or town) or city level; for example, the city of Berlin can be covered by a single TerraSAR-X StripMap scene [up to 3 m resolution, with a scene size of 30 km (width) × 50 km (length)]. This method is suitable for most cities and sites that are vulnerable to disasters. For example, Pakistan and India have military check posts on Siachen Glacier (located in the eastern Karakoram range in the Himalaya Mountains), where landslides and avalanches are quite frequent. A recent tragedy occurred in the Gyari sector of Pakistan in which 139 soldiers and civilians were buried under 80 feet of snow, when a huge avalanche hit a military base at the foot of Siachen glacier. The search and rescue teams faced difficulties in locating the exact site of the base camp; owing to the cloudy weather of the mountainous region, the acquisition of optical imagery is not always possible. Because of the huge amount of debris and snow, a more precise digging process was required. In such a situation, our proposed methodology could overcome these limitations. Similarly, in an urban environment, the proposed technique could be installed and used for the precise identification of potentially damaged areas and the estimation of the scale of damage to allow more precise and targeted search and rescue operations.

By using SAR remote sensing following a disaster event, we can precisely find the exact location ($L$), and then by using the semantic information from an existing 3D virtual model, we can plan more precise and efficient search and rescue operations. Figure 6 describes the possible schema for post-disaster planning using the semantic information from a 3D model. For example, in such a case, tunnel $T^h_d$ from the west with diameter $d$ and height $h$ would be the best choice because there is a high probability of finding bodies in area ($A$) near or around the main entrance (door).

## 7 Conclusion and Future Work

We have proposed a new model for rapid post disaster management and emergency response. Our proposed approach has good potential to meet future needs in the field of disaster management as a decision support system. By focusing on the current developments in the fields of science and technology, we can imagine how cities could be managed and organized in the future. We have proposed a new model for monitoring urban areas for both positive (new construction) and negative (building destruction) changes by using SAR data in combination with a 3D city

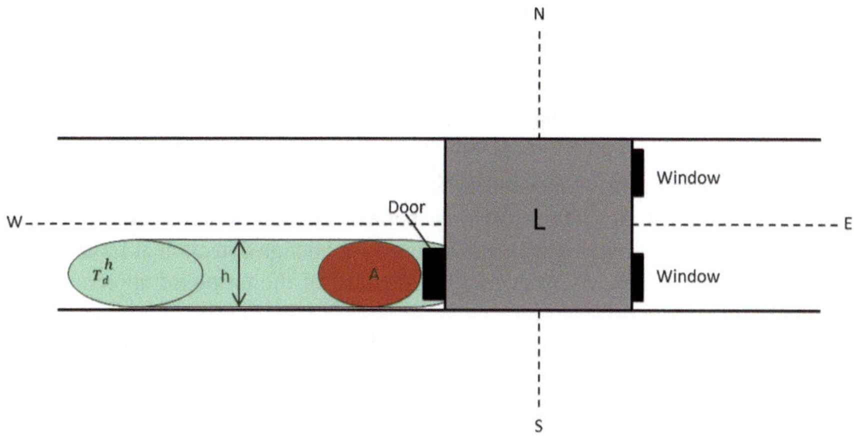


**Fig. 6** Planning framework based on semantic 3D building model

model database. We have discussed how semantic information from 3D city models can be used effectively for rapid post disaster management activities. To manage the uncertainties and to predict the scale of damage, the application of *interval-valued fuzzy soft sets* is implemented. The initial results show that the proposed model is highly feasibility and effective. In the future, the proposed model could be effectively used at a large scale for relief and rescue activities in modern cities.

Despite the fact that $\sigma^0_{[dB]post}$ will decrease for damage buildings as compare to the non-damage buildings (Matsuoka and Yamazaki 2004a, b), but still there are some other issues related to SAR geometric distortions (shadow, layover) and backscattering behavior are very important that need to be handled carefully. Specular scattering from the flat roof tops (multiple adjacent buildings) and could

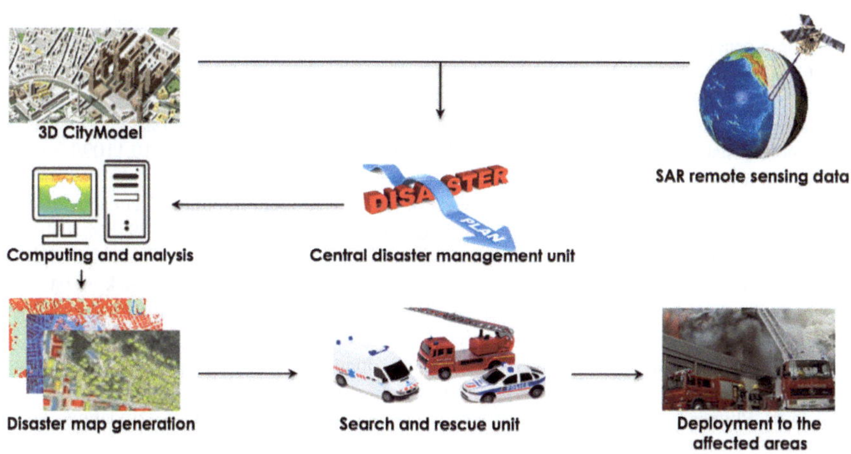


**Fig. 7** Graphical abstract of the proposed methodology

cause a lambertian scattering from the collapsed building and could create a relatively high backscattering as compare to the flat roof surface (Matsuoka and Yamazaki 2004a, b). The use of multi-sensor remote sensing can be used for the identification of damaged areas (Wang and Jin 2012). For level $L_2$ any remote sensing data and change detection algorithm can be used and the rule ($\sigma^0_{[dB]pre} > \sigma^0_{[dB]post}$) can be redefined accordingly. After going through the work done by Brunner et al. (2009, 2010) we believe that his approach for earthquake damage assessment is most feasible to be used for change detection at level $L_2$. The Fig. 7 shows the graphical overview (big picture) of the proposed approach.

**Acknowledgments** The authors would like to thank DigitalGlobe, Astrium Services, and USGS for providing the remote sensing data used in this study and the IEEE GRSS Data Fusion Technical Committee for organizing the Data Fusion Contest.

## References

Brunner D (2009) Advanced methods for building information extraction from very high resolution SAR data to support emergency response. Ph.D., University of Trento, Trento, Italy

Brunner D, Lemoine G, Bruzzone L (2010) Earthquake damage assessment of buildings using VHR optical and SAR imagery. IEEE Trans Geosci Remote Sens 48:2403–2420. doi:10.1109/TGRS.2009.2038274

Chetia B, Das PK (2010) An application of interval-valued fuzzy soft sets in medical diagnosis. Int J Contemp Math Sci 5:1887–1894

CityGML (2014) http://www.citygml.org. Accessed 4.24.14

Dekker PL (2010) Introduction to SAR [WWW Document]. https://earth.esa.int/web/guest/training-packages/-/article/esa-polish-academy-of-science-dlr-radar-remote-sensing-course-warsaw. Accessed 7.25.14

Dell'Acqua F, Gamba P (2012) Remote sensing and earthquake damage assessment: experiences, limits, and perspectives. Proc IEEE 100:2876–2890. doi:10.1109/JPROC.2012.2196404

Dell'Acqua F, Gamba P, Polli D (2010) Mapping earthquake damage in VHR radar images of human settlements: preliminary results on the 6th April 2009, Italy case. In: Geoscience and remote sensing symposium (IGARSS) 2010 IEEE International, pp 1347–1350. doi:10.1109/IGARSS.2010.5653973

Dell'Acqua F, Lisini G, Gamba P (2009) Experiences in optical and SAR imagery analysis for damage assessment in the Wuhan, may 2008 earthquake. In: IEEE international geoscience and remote sensing symposium 2009, IGARSS 2009, pp IV–37, IV–40. doi:10.1109/IGARSS.2009.5417603

Dong L, Shan J (2013) A comprehensive review of earthquake-induced building damage detection with remote sensing techniques. ISPRS J Photogramm Remote Sens 84:85–99. doi:10.1016/j.isprsjprs.2013.06.011

Feng F, Li Y, Leoreanu-Fotea V (2010) Application of level soft sets in decision making based on interval-valued fuzzy soft sets. Comput Math Appl 60:1756–1767. doi:10.1016/j.camwa.2010.07.006

Gröger G, Kolbe T, Czerwinski A, Nagel C (2008) Open GIS city geography markup language (CityGML) encoding standard (OGC 08-007r1)

Gröger G, Plümer L (2012) Provably correct and complete transaction rules for updating 3D city models. GeoInformatica 16:131–164. doi:10.1007/s10707-011-0127-6

Kagan YY (1997) Are earthquakes predictable? Geophys J Int 131:505–525. doi:10.1111/j.1365-246X.1997.tb06595.x

Kolbe TH (2009) Representing and exchanging 3D city models with city GML. In: Lee J, Zlatanova S (eds) 3D Geo-information sciences. Lecture notes in geoinformation and cartography. Springer, Berlin Heidelberg, pp 15–31

Kolbe TH, Gröger G, Plümer L (2005) CityGML: interoperable access to 3D city models. In: van Oosterom PDP, Zlatanova DS, Fendel EM (eds) Geo-information for disaster management. Springer, Berlin, Heidelberg, pp 883–899

Kubal C, Haase D, Meyer V, Scheuer S (2009) Integrated urban flood risk assessment—adapting a multicriteria approach to a city. Nat Hazards Earth Syst Sci 9:1881–1895. doi:10.5194/nhess-9-1881-2009

Lu D, Mausel P, Brondízio E, Moran E (2004) Change detection techniques. Int J Remote Sens 25:2365–2401. doi:10.1080/0143116031000139863

Lu W, Doihara T, Matsumoto Y (1998) Detection of building changes by integration of aerial imageries and digital maps. Int Arch Photogram Remote Sens 32:244–347

Maji PK, Biswas R, Roy AR (2001) Fuzzy soft sets. Fuzzy Math 9:589–602

Maji PK, Roy AR, Biswas R (2002) An application of soft sets in a decision making problem. Comput Math Appl 44:1077–1083. doi:10.1016/S0898-1221(02)00216-X

Martin R, Bernhard H (2010) Change detection of building footprints from airborne laser scanning acquired in short time intervals. Presented at the ISPRS commission VII mid-term symposium 100 Years ISPRS—advancing remote sensing science, Vienna, Austria, pp 475–480

Matsuoka M, Yamazaki F (2004a) Use of satellite SAR intensity imagery for detecting building areas damaged due to earthquakes. Earthq Spectra 20:975–994. doi:10.1193/1.1774182

Matsuoka M, Yamazaki F (2004) Building damage detection using satellite SAR intensity images for the 2003 Algeria and Iran earthquakes. In:IEEE international geoscience and remote sensing symposium 2004, IGARSS '04, pp. 1099–1102. doi:10.1109/IGARSS.2009.5417603

Molodtsov D (1999) Soft set theory—first results. Comput Math Appl 37:19–31. doi:10.1016/S0898-1221(99)00056-5

Murakami H, Nakagawa K, Hasegawa H, Shibata T, Iwanami E (1999) Change detection of buildings using an airborne laser scanner. ISPRS J Photogramm Remote Sens 54:148–152. doi:10.1016/S0924-2716(99)00006-4

Park S-E, Yamaguchi Y, Kim D (2013) Polarimetric SAR remote sensing of the 2011 Tohoku earthquake using ALOS/PALSAR. Remote Sens Environ 132:212–220. doi:10.1016/j.rse.2013.01.018

Suga Y, Takeuchi S, Oguro Y, Chen AJ, Ogawa M, Konishi T, Yonezawa C (2001) Application of ERS-2/SAR data for the 1999 Taiwan earthquake. Adv Space Res 28:155–163. doi:10.1016/S0273-1177(01)00334-9

Tong X, Lin X, Feng T, Xie H, Liu S, Hong Z, Chen P (2013) Use of shadows for detection of earthquake-induced collapsed buildings in high-resolution satellite imagery. ISPRS J Photogramm Remote Sens 79:53–67. doi:10.1016/j.isprsjprs.2013.01.012

Trianni G, Gamba P (2009) Fast damage mapping in case of earthquakes using multitemporal SAR data. J Real-Time Image Process 4:195–203. doi:10.1007/s11554-008-0108-7

Vogtle T, Steinle E (2004) Detection and recognition of changes in building geometry derived from multitemporal laserscanning data. Presented at the International Archives of Photogrammetry, Remote Sens Spat Inf Sci, Istanbul, Turkey, pp 428–433

Wang T-L, Jin Y-Q (2012) Postearthquake building damage assessment using multi-mutual information from pre-event optical image and postevent SAR image. IEEE Geosci Remote Sens Lett 9:452–456. doi:10.1109/LGRS.2011.2170657

Yang X, Lin TY, Yang J, Li Y, Yu D (2009) Combination of interval-valued fuzzy set and soft set. Comput Math Appl 58:521–527. doi:10.1016/j.camwa.2009.04.019

# Change Detection in CityGML Documents

**Richard Redweik and Thomas Becker**

**Abstract** With the increasing importance and usage of virtual 3D city models, a suitable continuation process becomes necessary to ensure the sustainability of 3D city models. If a new model was gathered e.g. by processing from LIDAR data or new data acquisition—knowledge about the differences between the old and the new city model is beneficial. This would allow the versioning, update and historicization. In this contribution we will present a concept for the detection of changes between two CityGML instance documents to ensure sustainable continuation processes in the field of 3D city models. The proposed concept will be based on techniques of existing change detection algorithms and object detection algorithms from the field of image analysis. The detected changes will be saved in a well-defined format so that following applications such as database operations can be performed or can make use of them. Finally, we examine the performance of our proof-of-concept implementation.

**Keywords** CityGML · 3D city model · Change detection · Historicization · Hough transform

## 1 Introduction and Motivation

The importance and use of virtual 3D city models is increasing nowadays (Gröger and Plümer 2012). Since "[…] cities change constantly and thus the models will be outdated one day" (Navratil et al. 2010), 3D city models need to be maintainable. Hence, a feasible update process is needed. The Open Geospatial Consortium (OGC) standard for the representation, storage and exchange of virtual 3D city

R. Redweik · T. Becker (✉)
Institute for Geodesy and Geoinformation, Technische Universität Berlin,
Straße des 17. Juni, 10623 Berlin, Germany
e-mail: thomas.becker@tu-berlin.de

R. Redweik
e-mail: richard.redweik@posteo.de

© Springer International Publishing Switzerland 2015 
M. Breunig et al. (eds.), *3D Geoinformation Science*,
Lecture Notes in Geoinformation and Cartography,
DOI 10.1007/978-3-319-12181-9_7

models is CityGML. The current workflow to update a 3D city model is to overwrite the complete database with the whole edited resp. created CityGML file, which is not very straight forward and sustainable.

If a new city model was gathered by surveying and the database is overridden completely by this new model, there is no chance to compare the continuously done changes against the surveying. In worst case, a change was done after the surveying for the complete city model. This means, actual data will be overwritten by outdated data from the survey.

To avoid such situations it would be useful to validate the changes before committing them to the RDBMS. Thus, only the changes may be written into the RDBMS. In addition, versioning and historicization can be enabled. To make a step towards a feasible update concept and thus towards a versioning and historicization concept on a technical level a diff-tool for CityGML would be beneficial.

To work with large city models efficiently, a 3D geo database like the 3D City Database (3DCityDB), which is implemented on top of a spatial relational database management system (RDBMS), can be used. The 3DCityDB is realized as a relational database schema for Oracle Spatial and for PostgreSQL/PostGIS. However, only Oracle Spatial implements a versioning mechanism (Gröger et al. 2005; Kunde 2012). Thus, PostgreSQL/PostGIS do not allow versioning and historicization. In addition, the demand to provide fundamental data for spatial referencing in a digital way is increasing for surveying, mapping and cadastral authorities. This includes 3D geo-information as well. Thus, to apply the versioning and historicization concept to 3D data a feasible update concept for CityGML based 3D city models is needed.

This paper addresses the research task on how to calculate the changes between two CityGML documents. We propose an approach to detect changes on the semantical and on the geometrical level. The main ideas are the application of concepts of existing change detection algorithms as well as concepts of object detection from the field of image analysis. The remaining paper is structured as follows. First, we introduce relevant related work, namely the concepts of CityGML and change detection algorithms. This is followed by a description on how to apply the introduced concepts for change detection in CityGML instance documents. These concepts are extended by an approach to match geographic features by their geometric characteristics. Furthermore, we introduce an exchange format that describes the detected changes. In the end the performance of our proof-of-concept implementation will be evaluated.

## 2 Related Work

### *2.1 CityGML*

CityGML is the Open Geospatial Consortium (OGC) standard for the representation, storage and exchange of virtual 3D city models. It is an application schema of the Geography Markup Language 3 (GML3) and thus based on Extensible Markup

Language (XML) (Gröger et al. 2012). XML represents documents as hierarchically structured information, which can be represented as a tree after the Document Object Model (DOM) specification (Wood et al. 2000). Unlike many other 3D vector formats which are based on geometry and appearance information, CityGML represents semantic and topological aspects of features which are relevant for city models as well. By providing a semantic "geospatial information model (ontology) for urban landscapes [...]" (Gröger et al. 2012), it cannot only be used for visualization purposes but also for thematic analysis. CityGML defines classes with its attributes and relations for the most relevant features for the application of 3D city models like buildings, vegetation, water bodies etc. The features are organized into modules: one core and several thematic extension modules (building, tunnel, bridge, vegetation ...). Since CityGML is a common information model, it can be used by various application fields, which again are able to define application specific extensions via Application Domain Extensions (ADE) (Gröger et al. 2012).

Spatial properties are represented by objects of a subset of the GML3 geometry model, a so called profile. The model represents 3D geometry according to the Boundary Representation (B-Rep). The B-Rep describes "[...] an object in terms of its surface boundaries: vertices, edges, and faces" (Foley et al. 1997, 542). Furthermore, the geometry model consists of primitives, which may be combined to form aggregates, complexes and composite geometries. For the modeling of topology CityGML does not use GML3's topology model but the XML Linking Language (XLink). Geometry objects are defined only once and referenced by features or by more complex geometries which have a topological relationship with this geometry. The referencing is done via Uniform Resource Identifiers (URI). CityGML represents features in five Levels of Detail (LOD), where the same object can be represented in different LODs simultaneously. LODs describe not only geometrical but also semantical aspects. With an increasing LOD-level objects become more accurate and detailed.

As Stadler et al. (2008) describe "[...] GMLIDs are unique in single instance documents only". Thus, it cannot be assumed that the ids stay constant between two versions of a CityGML document. Hence, different *gml:ids* of two features do not imply that these two features are distinct.

## *2.2 Change Detection Algorithms*

As Chawathe et al. (1996) point out "detecting changes to data [...] is a basic function of many important database facilities and applications [...]". The GNU *diff* utility written by Hunt and McIlroy (1976) is a famous utility to compute differences between flat files. However, it is not able to "understand the hierarchical structure of data" (Chawathe et al. 1996) as given in CityGML. Thus, *diff* is not suitable for change detection in CityGML documents. Though, there exist algorithms for change detection in hierarchically structured information like XML

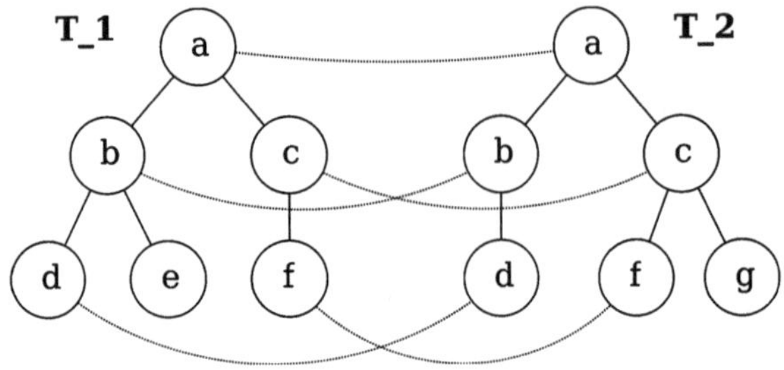


**Fig. 1** A partial matching between the trees $T_1$ and $T_2$. The matches are visualized by *dashed lines*

documents. The basic idea of change detection algorithms for hierarchically structured information is to calculate a set of edit operations which transform one tree into the other.

Essentially, change detection algorithms for hierarchically structured information work in several phases. First, corresponding nodes need to be found in the two trees $T_1$ and $T_2$ whose values are unchanged or updated. The correspondence of nodes is called a matching. A matching is called total, if all nodes of the two trees can be matched, if not, the matching is partial. Figure 1 visualizes a partial matching between two trees. If the elements contain object identifiers, matching is relatively easy (Chawathe et al. 1996). Next a sequence of change operations (e.g. insert, delete, update) which transform the tree $T_1$ into a tree $T_1'$ which is isomorphic to $T_2$, need to be found. Thus, if a partial matching M between $T_1$ and $T_2$ is found (see Fig. 1) and $T_1$ is transformed into $T_1'$, "[…] the partial matching M is extended into a total matching M between the nodes of $T_1$ and $T_2$" (Chawathe et al. 1996). Then M defines the isomorphism between $T_1$ and $T_2$. Such a sequence of edit operations is called an edit script (see Fig. 2).

Basically, algorithms for change detection on trees can be divided into two categories: one that processes on ordered and one that processes on unordered trees. In an unordered tree only the ancestor relationship of nodes and not the left-to-right

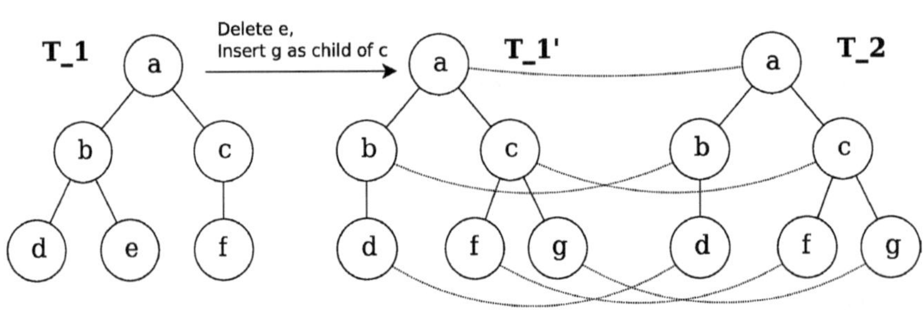


**Fig. 2** Extending a partial matching into a total matching between $T_1'$ and $T_2$. The edit operations *Delete e* and *Insert g as a child of c* transform $T_1$ into $T_1'$ which is isomorphic to $T_2$

ordering of siblings is significant as in CityGML. If for instance only the ordering of buildings in the document changed, it still describes the same area as before. Semantically the document did not change.

An algorithm for change detection in XML documents represented by unordered trees is X-Diff developed by Wang et al. (2003). It uses several concepts which can be exploited for change detection in CityGML documents. However, since *gml:ids* may change between two versions of a city model, an id-based matching approach like X-Diff will fail in this situations. Thus, this approach will be extended by an approach exploiting geometrical characteristics of the data.

## 3 Changes in CityGML Documents

Several change operations between two different versions of a CityGML document are possible. However, we assume that all changes are compliant to the CityGML schema. First of all, elements, which can be a feature or a property, may be inserted or deleted. In this case all of the element's child elements need to be inserted resp. deleted, too. The child elements may be specified by-value or by-reference (via XLink). Note that our approach does not resolve XLinks. If the referenced value changed, but the reference (XLink) did not, the change will be detected and processed at the location of this value. However, if a feature, which is referenced by XLink is deleted, the XLink needs to be deleted, too. Furthermore, the value of an element can be updated. For instance, this may be the change of the value of a generic attribute or the change of a geometry's coordinates. If an element was updated all its parent elements are considered to be updated as well. In addition, an element's attribute which is defined by a key-value pair may be inserted or deleted, or its value may be updated. Since it cannot be assumed that *gml:ids* stay constant between two versions of a CityGML document, matching by id may fail. However, the spatial properties of geometrical features can be utilized to match them. Furthermore, the CityGML documents may have different sources and thus may be modeled with different accuracies. This means the objects may be modeled in different LOD categories.

## 4 Change Detection

In this section we describe how to detect changes between two CityGML documents. For this task we make use of concepts from existing change detection algorithms (cf. *X-Diff* developed by Wang et al. 2003). Note that the methods presented here are mainly focused on the core and the building module. As Gröger and Plümer (2012) point out "the building model is the most important component of CityGML". Thus, this is a good starting point for the development of a tool for change detection in CityGML documents.

## *4.1 Tree Representation of CityGML Documents*

Since CityGML is based on XML, the documents can be represented as trees. In this research we represent them as modified and simplified Document Object Model (DOM) trees. Figure 3 visualizes our tree structure. The abstract base type *AbstractTreeNode* holds references to its parent node and its children nodes. Furthermore, it holds a signature and a hash value. *ElementNode* is a specialization of *AbstractTreeNode* and represents an element in a CityGML document. Beyond the inherited attributes it has a name, which is the value of the CityGML element's name (e.g. "building" for a building), and a value, which is the CityGML element. *AbstractLeafNode* is the abstract base class for leaf nodes. It holds a value, which is a string representing the leaf node's value. *AttributeNode* represents the attribute of an element with the attribute *attributeName*. *TextNode* represents a leaf-node's value as a string.

The signature is utilized to find matches between nodes in the two trees. It serves as a matching criterion, similar to Wang et al. (2003). The signature of a node is the concatenation of the names and (if existent) the *gml:ids* of its ancestors, similar to XPath. Additionally, the node's own name and *gml:id* (if existing) is added to the signature. Since text nodes do not have a name, the type is added to the signature. Thus, if x is a node in a tree T, then its signature is

$$\mathrm{Signature}(x) = /\mathrm{Name}(x_1)/\mathrm{gml{:}id}(x_1)/\ldots/\mathrm{Name}(x_n)/\mathrm{gml{:}id}(x_n)/\mathrm{Name}(x)/\mathrm{gml{:}id}(x).$$

Note that the *gml:id* is added to the signature only if the element is a specialization of *AbstractGML* and thus has a *gml:id*. Utilizing the signature prevents false matching of nodes with the same name but different context (e.g. the creation date of a building and the creation date of a bridge).

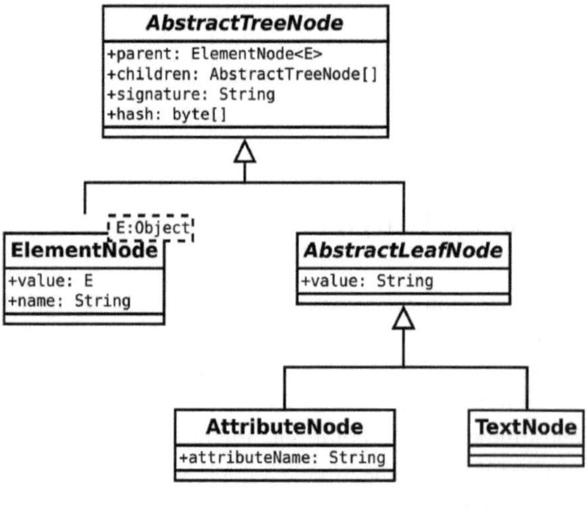


**Fig. 3** UML class diagram visualizing the defined tree structure

The hash value of a node is a representation of the subtree with its data rooted at that node. It is calculated by a hash function, which is chosen such that it is infeasible to find distinct input data sets with the same hash value. The hash value is like a "fingerprint" of the input data. If the data changed, the hash value changed as well. Thus, two subtrees with equal hash values are considered as being equal. Hence, the hash values can be used to find subtrees in the two documents that did not change.

### 4.2 Edit Operations and Edit Script

The definition of the edit operations is similar to the one used by Wang et al. (2003) and Chawathe et al. (1996). $e$ is an edit operation if the tree $T_2$ is the result of applying $e$ to the tree $T_1$: $T_1 \rightarrow T_2$.

Basically, there are three edit operations:

- *Insert*(*x*, *y*):
  Insert a node x as a child of node y. The children of y do not become children but siblings of x.
- *Delete*(*x*):
  Delete the node x from the tree.
- *Update*(*x*, *value*):
  Update the value of the *AbstractLeafNode* x with the specified value.

Since the CityGML document is represented as an unordered tree, the position to insert a node x as a child of node y does not matter.

A sequence of edit operations which transform one tree into the other is called an edit script. It is defined as: $E = e_1, \ldots, e_m$ is an edit script, if there are trees $T_2, \ldots, T_m$ such that $T_1 \xrightarrow{e_1} T_2 \xrightarrow{e_2} \cdots \xrightarrow{e_m} T_{m+1} : E(T_1 \rightarrow T_{m+1})$. The task in this research is to find an edit script $E(T_1 \rightarrow T_1')$ that transforms the tree $T_1$ into a tree $T_1'$ which is isomorphic to $T_2$.

### 4.3 Matching

As stated earlier a matching is the correspondence of nodes in the two trees. More formally, a set of node pairs (*x*, *y*) is a matching M from the tree $T_1$ to tree $T_2$, iff

1. (a) $(x, y) \in M$, $x \in T_1$, $y \in T_2$, $Signature(x) = Signature(y)$, or
   (b) $(x, y) \in M$, $x \in T_1$, $y \in T_2$, the geometry of x and y is matched spatially,
2. The nodes x and y are only matched, if their ancestors x′ and y′ holding an object which inherits from *AbstractGML* are matched. More formally, if $(x, y) \in$

$M$, and if x′ is an ancestor of x and y′ is an ancestor of y and x and y holding an object of type *AbstractGML* as the value, then $(x', y') \in M$.
3. Matches are one-to-one. That is $\forall(x, y) \in M$ and $(x', y') \in M$, $x = x'$ *iff* $y = y'$.

The subdivision of the first constraint enables not only to match GML3 objects with the same *gml:id* and thus signature, but also to match spatial objects which have the same location and extent but different *gml:ids*.

The second criterion prevents matching of elements of different features. For example, the creation date of building A must not be matched with the creation date of building B. This influences the way how to handle a change of geometries in which the underlying geometry elements are reused. For instance, the geometry of a building may change from a solid to a multi-surface or vice versa and the surface-members are reused. Since the reused geometry elements are not matched, the algorithm detects the change as a deletion of the old geometry followed by an insertion of the new geometry. Trying to match the underlying geometry elements would increase the algorithm's run-time complexity. Thus, it seems to be justifiable to replace a move operation of the underlying geometry elements by a sequence of delete and insert statements of the whole geometry element.

**From Matching to Edit Script** Based on the matches M from tree $T_1$ to tree $T_2$ the edit script can be calculated. Inserts are nodes in the tree $T_2$ which are not in M. Vice versa, deletions are calculated based on nodes which are in tree $T_1$ but not in M. Updates describe matched nodes with different values.

# 5 Spatial Matching

The first matching criterion of the previous section is divided into the two sub-criteria of matching by signature and by spatial properties. Latter enables the matching of spatial objects with different *gml:ids* which represent the same real-world object. These objects have the same spatial location but different signatures. This section describes the calculation of spatial matches. Since the research focuses on the building module, the methods for spatial matching are developed mainly for this module. However, those may be applied to other modules such as bridge or tunnel. Our approach is to match top-level features (buildings) first, followed by matching their surface geometries.

## *5.1 Matching Top-Level Features*

Kolbe et al. (2009) present a 2D approach to find corresponding building models. Their main idea is to compare the 2D footprints of the buildings by neglecting the height information. Two buildings are considered as matching if their footprints

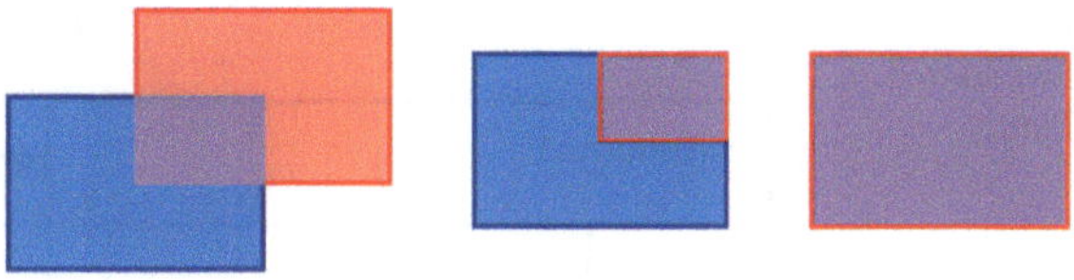


**Fig. 4** Overlap of two objects defined by two percentage values (adapted from Kolbe et al. 2009)

overlap to some degree. Therefore the overlapping percentages for both buildings need to be found (see Fig. 4):

$$((area_{ref} \cap area_{merge})/area_{ref}) \text{ and } ((area_{merge} \cap area_{ref})/area_{merge}) \qquad (1)$$

We extended this approach by two parameters. First, the classes of the two top-level features need to match. Second, the top-level features' heights need to overlap to some degree, similar to the given formula, but with the height values referenced linearly. Our extensions prevent the matching of features of different classes (e.g. building and tunnel) and the matching of features of the same class modeled above each other (e.g. a motorway bridge modeled above another motorway bridge).

## *5.2 Matching Surface Geometries*

The drawback of the footprint and height approach for spatial matching is that it is applicable for objects with a great extent in 2D only. If there is a shift between the two city models (e.g. due to transformation errors or changed accuracy) the approach may match the buildings. However, it is not able to match wall surfaces, since there may be no overlap of their 2D footprints. Figure 5 visualizes a situation of two shifted and matched buildings. The two walls *Wall1* and *Wall2* do not overlap. Thus, to match surface geometries of already matched top-level features a 3D Hough transform will be utilized. The Hough Transform is a method for detecting parameterized objects. It was invented by Hough (1962) to detect lines in pictures. The generalized version of the Hough transform can be used to detect parameterized objects.

We are utilizing a 3D Hough transform as follows. As the input to the transform we use the surface geometries of two matched top-level features. Our parameter space consists of three parameters. The first one is the length of a vector $\overrightarrow{r}$ pointing from the surface's reference point to the building's reference point. The reference point of a feature may be its centroid or *lowerCorner*. The second parameter is the surface geometry's normal vector $\overrightarrow{n}$. The third parameter is the surface's semantics. The semantics is made up of the surface's LOD and its relation to the building.

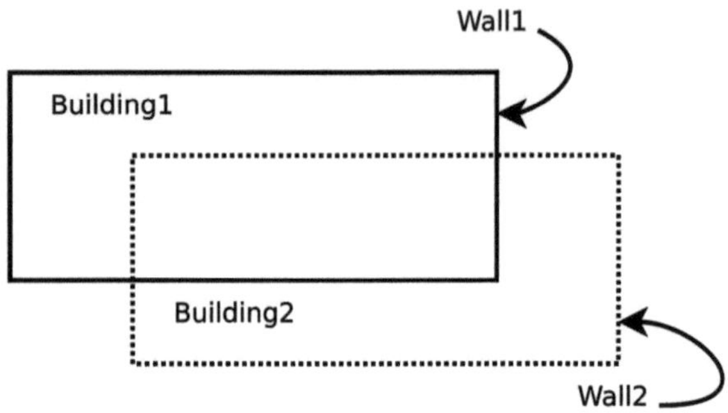


**Fig. 5** Schematic visualization of two shifted but matched buildings

The usage of the LOD is necessary to prevent the matching of surfaces of different LODs. Furthermore, it enables to detect deletions and insertions of surfaces in different LODs. After Gröger et al. (2012) the relation of the surface to the building can be *lod*Solid*, *lod*MultiSurface* or *boundedBy* (* describes the LOD-level: 1–4). In case of *boundedBy* the relation can be differentiated semantically by its function like *RoofSurface*, *WallSurface*, *GroundSurface* and more. To enable the matching of slightly rotated surfaces the Hough space needs to be extended by a rotation angle for the normal vector $\vec{n}.\vec{r}$ and thus $r = |\vec{r}|$ may change as well if the surface is rotated. However, the rotation and variation of r shall happen only in a very small range to adjust small misalignments or small rotations (e.g. due to CRS changes). Two surfaces with the same parameterization are matching. Hence, we are not building an accumulator array in which we are looking for maxima of votes, but looking for two surfaces within an ε-range.

Thus, this procedure allows to match the surfaces with changed *gml:ids* of matched top-level features invariant to rotations and shifts of the top-level features. Surfaces which are rotated and shifted in a small range can be matched, too. Hence, this method allows the matching of the two wall surfaces *Wall1* and *Wall2* shown in Fig. 5. In addition, if a matching between two features with different *gml:ids* is found, possible XLink-references to this feature need to be updated as well.

Furthermore, this approach allows detecting changes between documents from different sources with objects modelled in different LODs. Top-level features modelled in different LODs can be matched by their footprint and height overlaps, which is invariant to LOD categories. However, surface geometries modelled in different LODs cannot be matched, since the matching procedure considers the surfaces' semantics which includes the LOD. Thus, surface geometries modelled in different LODs are handled by a deletion and an insertion.

However, even if this procedure is developed mainly for the building module it can be applied to other modules as well. These modules need to specify the elements' geometries by surfaces. These modules are amongst others the tunnel and bridge module. Though, the 3D Hough transform cannot be applied to geometry elements specified prototypically by an implicit geometry.

# 6 Change Detection Algorithm

This section describes the work-flow of the algorithm in detail as defined by the modules described in the previous sections.

First the two CityGML documents which shall be compared need to be parsed and validated so that the following modules have access to the instantiated objects. The proof-of-concept uses the Java class library and API citygml4j for the reading of CityGML documents. Citygml4j parses and validates the two documents and offers access to the instantiated features, which will be processed in the next steps.

Afterwards the object graph given by citygml4j is transformed into a tree structure as described by Fig. 3. This is necessary since the graph structure as defined by the CityGML schema does not comply with the structure needed by the following modules.

After the two trees are built up, the matches between them are calculated. First the nodes are matched by signature. If there are nodes which cannot be matched by signature, those are matched by their spatial properties:

- *Matching by Signature*: Nodes which have the same signature and the same hash value are removed from the trees. These nodes did not change between the two versions of the CityGML document. The removing of unchanged nodes reduces the search space for finding the matches in the next step. Next the matches between two trees are calculated in a top-down approach. Two nodes are matching if their signature, class and id are equal. Two matching nodes are added to the distance table, which holds the matches.
- *Spatial Matching*: If not all nodes in the two trees are matched by signature, the algorithm tries to match these nodes by their spatial properties. This is done as described in Sect. 5.

If the matches are calculated, their distances will be computed in a bottom-up way. The distance between two leaf-nodes is 0 if the values are equal, 1 otherwise (update, insert, delete). The distance of non-leaf nodes is defined as the sum of their children's distances. The calculated distance is assigned to the corresponding matching in the distance table. In a last step the edit operations are determined. The detected edit operations are collected in a list, which makes up the edit script.

# 7 Exchange Format for Detected Changes

To enable the use of the detected changes for following applications, it is necessary to store these changes in a well-defined exchange format. This section describes such a format.

Our edit script is defined by a XML Schema (Fig. 6). Thus, the exchange format is XML based. It defines the elements *InsertOperation*, *DeleteOperation* and *UpdateOperation* with their specific types, so that the exchange format is able to describe the three edit operations (insert, delete, update). The location of the changes is given by XML Path Language (XPath). The specification of the changes is done by specifying the inserted or deleted element respectively the old and new value of the updated element in-line. In case of deletes, this produces a slight overhead in the size of the edit script on the one hand, but on the other hand ensures that a third party application (e.g. for visualization purposes) knows exactly what was changed without considering the original CityGML documents and additional processing. Thus, an application using this edit script can reconstruct and make use of the changes without considering the original or changed CityGML document. Hence, the exchange format enables the creation of database commits or WFS-T statements for the changes only and thus enables versioning and historicization for existing databases.

```xml
<?xml version="1.0" encoding="UTF-8"?>
<xs:schema xmlns:xs="http://www.w3.org/2001/XMLSchema"
           xmlns="http://igg.tu-berlin.de"
           targetNamespace="http://igg.tu-berlin.de"
           elementFormDefault="qualified">
    <xs:element name="EditScript" type="EditScriptType"/>
    <xs:complexType name="EditScriptType">
        <xs:sequence>
            <xs:element name="editOperation" type="EditOperationType" maxOccurs="unbounded"/>
        </xs:sequence>
    </xs:complexType>
    <xs:element name="_EditOperation" type="EditOperationType" abstract="true"/>
    <xs:complexType name="EditOperationType" abstract="true">
        <xs:sequence>
            <xs:element name="xPath" type="xs:string"/>
        </xs:sequence>
    </xs:complexType>
    <xs:element name="UpdateOperation" type="UpdateOperationType" substitutionGroup="_EditOperation"/>
    <xs:complexType name="UpdateOperationType">
        <xs:complexContent>
            <xs:extension base="EditOperationType">
                <xs:sequence>
                    <xs:element name="oldValue" type="xs:anyType"/>
                    <xs:element name="newValue" type="xs:anyType"/>
                </xs:sequence>
            </xs:extension>
        </xs:complexContent>
    </xs:complexType>
    <xs:element name="_InsertDeleteOperation" type="InsertDeleteOperationType" abstract="true"
                substitutionGroup="_EditOperation"/>
    <xs:complexType name="InsertDeleteOperationType" abstract="true">
        <xs:complexContent>
            <xs:extension base="EditOperationType">
                <xs:sequence>
                    <xs:element name="value" type="xs:anyType"/>
                </xs:sequence>
            </xs:extension>
        </xs:complexContent>
    </xs:complexType>
    <xs:element name="InsertOperation" type="InsertOperationType" substitutionGroup="_InsertDeleteOperation"/>
    <xs:complexType name="InsertOperationType">
        <xs:complexContent>
            <xs:extension base="InsertDeleteOperationType"></xs:extension>
        </xs:complexContent>
    </xs:complexType>
    <xs:element name="DeleteOperation" type="DeleteOperationType" substitutionGroup="_InsertDeleteOperation"/>
    <xs:complexType name="DeleteOperationType">
        <xs:complexContent>
            <xs:extension base="InsertDeleteOperationType"></xs:extension>
        </xs:complexContent>
    </xs:complexType>
</xs:schema>
```

**Fig. 6** XML schema document of the exchange format

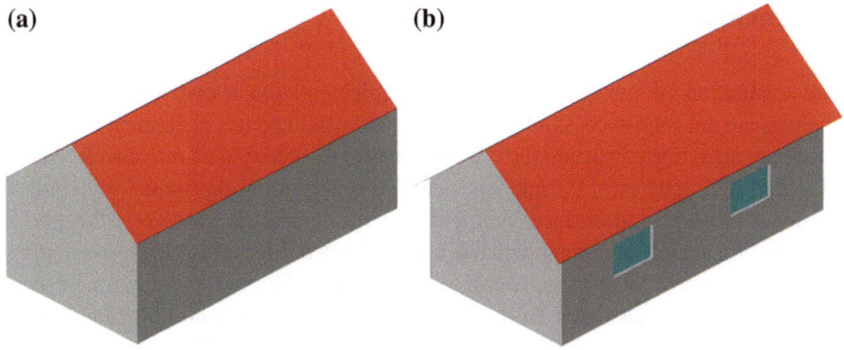


**Fig. 7** The two buildings used for difference calculation. **a** The building modeled in LOD2. **b** The building modeled in LOD3

```
<EditScript ...>
    ...
    <editOperation xmlns:xsi="http://www.w3.org/2001/XMLSchema-instance" xsi:type="UpdateOperationType">
        <xPath>/CityModel/cityObjectMember/Building[@gml:id="GML_7b1a5a6f-ddad-4c3d-a507-3eb9ee0a8e68"]
            [@gml:id]</xPath>
        <oldValue xsi:type="xs:string">GML_7b1a5a6f-ddad-4c3d-a507-3eb9ee0a8e68</oldValue>
        <newValue xsi:type="xs:string">GML_7b1a5a6f-ddad-4c3d-a507-3eb9ee0a8e68-LOD3</newValue>
    </editOperation>
     <editOperation xmlns:xsi="http://www.w3.org/2001/XMLSchema-instance" xsi:type="InsertOperationType">
     <editOperation xmlns:xsi="http://www.w3.org/2001/XMLSchema-instance" xsi:type="InsertOperationType">
     <editOperation xmlns:xsi="http://www.w3.org/2001/XMLSchema-instance" xsi:type="InsertOperationType">
        <xPath>/CityModel/cityObjectMember/Building[@gml:id="GML_7b1a5a6f-ddad-4c3d-a507-3eb9ee0a8e68-LOD3"]
            /boundedBy/GroundSurface</xPath>
        <value>
            <gml:name>Ground Slab</gml:name>
            <bldg:lod3MultiSurface>
                <gml:MultiSurface>
                    <gml:surfaceMember>
                        <gml:Polygon gml:id="GML_d3981803-d4b0-4b5b-969c-53f657594757">
                            <gml:exterior>
                                <gml:LinearRing>
                                    <gml:posList>458875.0 5438355.0 112.0 458885.0 5438355.0 112.0
                                        458885.0 5438350.0 112.0 458875.0 5438350.0 112.0
                                        458875.0 5438355.0 112.0</gml:posList>
                                </gml:LinearRing>
                            </gml:exterior>
                        </gml:Polygon>
                    </gml:surfaceMember>
                </gml:MultiSurface>
            </bldg:lod3MultiSurface>
        </value>
    </editOperation>
    ...
    <editOperation xmlns:xsi="http://www.w3.org/2001/XMLSchema-instance" xsi:type="DeleteOperationType">
        <xPath>/CityModel/cityObjectMember/Building[@gml:id="GML_7b1a5a6f-ddad-4c3d-a507-3eb9ee0a8e68"]
            /boundedBy/GroundSurface</xPath>
        <value>
            <gml:name>Ground Slab</gml:name>
            <bldg:lod2MultiSurface>
                <gml:MultiSurface>
                    <gml:surfaceMember>
                        <gml:Polygon gml:id="GML_d3981803-d4b0-4b5b-969c-53f657594757">
                            <gml:exterior>
                                <gml:LinearRing>
                                    <gml:posList>458875.0 5438350.0 112.0 458875.0 5438355.0 112.0
                                        458885.0 5438355.0 112.0 458885.0 5438350.0 112.0
                                        458875.0 5438350.0 112.0</gml:posList>
                                </gml:LinearRing>
                            </gml:exterior>
                        </gml:Polygon>
                    </gml:surfaceMember>
                </gml:MultiSurface>
            </bldg:lod2MultiSurface>
        </value>
    </editOperation>
    ...
</EditScript>
```

**Fig. 8** Excerpt from the edit script generated from the changes between the building modeled in LOD2 and LOD3

# 8 Conclusion

The evaluation of our proof-of-concept implementation shows that the defined changes are detected successfully. Inserts, deletes and updates of elements as well as attributes are detected successfully. Furthermore, the deletion of referenced features (XLinks) and updates of XLinks are detected. In addition, the spatial matching of top-level features and surface geometries works correctly. However, the performance of the spatial matching depends on the defined threshold values (overlapping percentage values, rotation angle and ε range-values). Figure 8 visualizes an excerpt of the edit script calculated from the differences between a building modeled in LOD2 and a building modeled in LOD3 (see Fig. 7). Both buildings are taken from the example datasets in the CityGML 2.0.0 XML Schema package (Building_LOD2-EPSG25832.gml, Building_LOD3-EPSG25832.gml). We changed the gml:ids of the LOD3 building to prevent the matching by signature. The edit script shows a sample insert, delete, and update operation. It can be seen that the polygon with *gml:id* = "*GML_d3981803-d4b0-4b5b-969c-53f657594757*" is deleted as a lod2MultiSurface but inserted as a lod3MultiSurface. In addition, the change of the building's *gml:id* is detected successfully.

Since the LOD change is detected correctly, our proof-of-concept is able to handle and match objects modeled in different LODs with changed *gml:ids* correctly. The buildings are matched by the footprint and height approach, while the surface geometries modeled in different LODs are not matched. Their change is represented by a deletion and an insertion.

In this paper, we have discussed how to detect changes between two different versions of a CityGML document. We examined how to apply approaches of existing change detection algorithms to CityGML. However, by the characteristics of CityGML, namely the possibility of changing *gml:ids* and the representation of spatial data, these approaches may fail to match elements. We proposed solutions to this by matching geometrical elements spatially. Top-Level features are matched by their 2D footprint and height information. Surface geometries of matched top-level features are matched by a 3D Hough transform by looking for surfaces with the same parameters. Our proof-of-concept implementation demonstrates that the defined changes are detected successfully, at least for small regions with sparse buildings. Furthermore, it is possible to match top-level features from different sources (different *gml:ids*) with changed quality (LOD). However, the surface geometries modelled in different LODs are not matched.

Future works needs to be done to extent our approach to enable the calculation of differences of all CityGML modules. In addition, the performance of applying our tool to whole cities needs to be evaluated. This includes the qualitative as well as run-time performance. Moreover, work needs to be done to detect topological changes between features.

## References

Chawathe SS, Rajaraman A, Garcia-Molina H, Widom J (1996) Change detection in hierarchically structured information. In: Proceedings of the 1996 ACM SIGMOD international conference on management of data (SIGMOD'96), Montreal, June 1996

Foley JD, van Dam A, Feiner SK, Hughes JF (1997) Computer graphics: principles and practice, 2nd edn. Addison-Wesley, Boston

Gröger G, Plümer L (2012) CityGML—Interoperable semantic 3D city models. ISPRS J Photogrammetry Remote Sens 71:12–33

Gröger G, Kolbe TH, Nagel C, Häfele K-H (2012) OpenGIS city geography markup language (CityGML) encoding standard, version 2.0.0. doc. no. 12-019. OGC, Wayland

Gröger G, Kolbe TH, Schmittwilken J, Stroh V, Plümer L (2005) Integrating versions, history and levels-of-detail within a 3D geodatabase. In: Proceedings of the 1st international ISPRS/EuroSDR/DGPF-workshop on next generation 3D city models. Bonn

Hough PVC (1962) Method and means for recognizing complex patterns. US Patent 3,069,654

Hunt JW, McIlroy MD (1976) An algorithm for differential file comparison. Technical report 41. AT&T Bell Laboratories, Murray Hill NJ

Kolbe TH, König G, Nagel C, Stadler A (2009) 3D-geo-database for CityGML. Version 2.0.1. Documentation. Technical report. http://www.3dcitydb.net/3dcitydb/fileadmin/downloaddata/3DCityDB-Documentation-v2_0.pdf. Accessed 12 June 2014

Kunde F (2012) CityGML in PostGIS: Portierung, Anwendung und Performanz-Analyse am Beispiel der 3DCityDB Berlin. Unpublished master thesis. Institut für Geographie, Universtät Potsdam, In German only. http://opus.kobv.de/ubp/volltexte/2013/6365. Accessed 12 June 2014

Navratil G, Bulbul R, Frank AU (2010) Maintainable 3D models of cities. In: Proceedings of the 15 international conference on Urban Planning, Regional Development and Information Society. Real CORP 2010, Vienna

Stadler A, Nagel C, König G, Kolbe TH (2008) Making interoperability persistent: a 3D geo database based on CityGML. In: Proceedings of the 3rd international workshop on 3D geo-information. Springer, Berlin, p 175

Wang Y, DeWitt DJ, Cai J-Y (2003) X-Diff: an effective change detection algorithm for XML documents. In: Proceedings of 19th international conference on data engineering, Bangalore

Wood L, Le Hors A, Apparao V, Byrne S, Champion M, Isaacs S, Jacobs I, Nicol G, Robie J, Sutor R, Wilson C (2000) Document object model (DOM) level 1 specification, 2nd edn. World Wide Web Consortium, Working Draft WD-DOM-Level-1-20000929, http://www.w3.org/TR/2000/WD-DOM-Level-1-20000929. Accessed 12 June 2014

# Change Detection of Cities

**F. Pédrinis, M. Morel and G. Gesquière**

**Abstract** Today, many cities have at their disposal a digital model useful in many applications such as decision making in urban planning. 3D data representing objects in the city such as land and buildings often comes from successive acquisition campaigns. Unfortunately, digital models of cities can have many versions of the same area. Having tools to detect changes becomes a necessity. It is then possible to highlight any differences between multiple versions of the same area in 3D. A second application can be related to the possibility of making a temporal representation by taking into account the detected changes. In this paper, we propose a set of tools to detect changes. The use case is done on buildings. Our method is based on CityGML and cadastre files. The output is a CityGML file containing a representation of the evolution over time of the objects in the city.

**Keywords** 3D city modeling · Change detection · Urban planning · CityGML · Temporality

## 1 Introduction

Over the last decade, the possibility of owning 3D data has become a reality in GIS. This data can be used in many applications such as decision making for urban planning or for the simulation of physical phenomena (sun exposure, pollution or flood propagation, etc.). However, as the city is changing rapidly over time, it is necessary to take into account these changes and develop a real lifecycle associated

F. Pédrinis · M. Morel · G. Gesquière (✉)
Commission II, WG II/2, LIRIS University of Lyon, Lyon, France
e-mail: gilles.gesquiere@liris.cnrs.fr

F. Pédrinis
e-mail: f.pedrinis@gmail.com

M. Morel
e-mail: maxime.morel@liris.cnrs.fr

© Springer International Publishing Switzerland 2015

M. Breunig et al. (eds.), *3D Geoinformation Science*,
Lecture Notes in Geoinformation and Cartography,
DOI 10.1007/978-3-319-12181-9_8

with this data, particularly for buildings in the context of this paper. A building can be constructed, modified and destroyed and then replaced by one or more others. These changes in the city over the years lead to consider how to take into account the representation in databases, but also when using exchange formats such as CityGML. One possibility to keep this temporal information related to the objects that compose the city was proposed by Morel (2014). This solution allows us to implement the management of several states of the same object of the city, each state being associated with temporal information. For example, an object can be a building, a bridge, a parcel of land or a texture associated with one of these objects. However, as explained by Morel in this paper, manual update cannot be considered because it would be time consuming. Users should change the model by processing each object of the city separately. It is then necessary to develop methods for automatic detection of changes in objects of the city over time. The goals are multiple. These methods would automatically update a dataset of the city by comparing it to new data. It will then be possible to detect changes between these two representations of the same scene as it would be the case with classic file versioning tools.

In the work presented in this paper, the data are CityGML files in LOD2, acquired in 2009 and 2012 and granted by the city of Lyon in France (Fig. 1, left). Potential links between the two sets of data cannot rely on the identifiers associated with each building, as they are randomly assigned during each acquisition campaign. The CityGML files contain buildings themselves composed of polygons listed as wall or roof. However, a set of buildings can be often grouped with neighbouring buildings in a single CityGML building element. For example, Fig. 1 (middle) shows in red highlighted an area (which has only one unique identifier) containing a large group of buildings. Modification of one of the buildings in this area should not lead to the change of the whole area. This dataset consisting of two CityGML files can be completed by the French cadastre in which we find building footprints (Fig. 1, right). However, it should be noted that these footprints are not obtained with the same acquisition protocols. It is therefore possible to find many differences between the CityGML data and cadastral data. Finally, aerial images which helped reconstructing 3D buildings are not always available; it is not possible

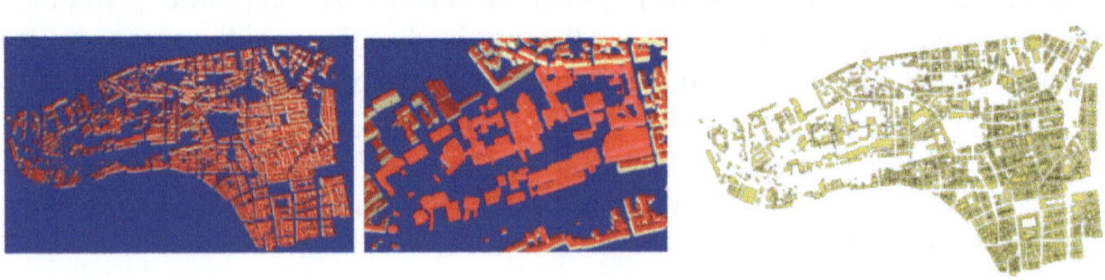

**Fig. 1** CityGML and cadastre files of Lyon 1. *Left* CityGML data. *Middle* the same building "id" may be composed of multiple buildings. *Right* cadastral information related to the same area

to compare them to detect changes. Anyway, these changes should have been carried on 3D models in the CityGML files. It is therefore necessary to establish a method for detecting changes based on CityGML and cadastral files.

The goal of this paper is to present a new method for detecting changes related to buildings in a city between two dates using CityGML files. Additionally, complementary cadastral data can be used if they are available. This can improve accuracy in detecting changes. This method is based on footprints computation of buildings. It is then possible to link additional items to 3D data that will help to detect changes related to the shape or height of the building. Cadastral information will provide relevant cutting of CityGML buildings.

This paper is organized into 8 parts. First we will overview the state of the art in Sect. 2. Then we will present our method to compare two CityGML files in Sect. 4 while having extracted footprints in a previous phase (Sect. 3). This method will be improved by taking into account the cadastre in Sect. 5. Section 6 will present results related to the area of Sathonay Camp and Villeurbanne near Lyon where we have two CityGML files. Finally, Sect. 7 demonstrates the relevance of using cadastral information. We then conclude and present some future works.

## 2 State of the Art

The goal of this paper is based on the comparison of CityGML data. In particular, in the context of this study, we are interested in the modification of buildings over time. This method should allow us to detect changes between two versions of CityGML data acquired in a same area at two different dates. A CityGML file gathering this information will be then generated as proposed in Morel (2014).

The study of the evolution of cities across their various models has already been the subject of numerous works in recent years. Different data sources can be used, as aerial imagery, satellite imagery, or LiDAR.

Thus, Champion compares Digital Surface Model (DSM) and Colour InfraRed Orthophotos (CIR) with an outdated vector database (Champion 2007). In his method, geometric primitives for buildings are extracted from DSM and CIR Orthophotos to be compared with the primitive of each object from the database. This comparison is made by using a similarity score to distinguish different cases: unchanged, altered or demolished building. Any new potential buildings are obtained from DSM and Digital Terrain Model (DTM) by extracting elements above ground that cannot be assimilated to vegetation and are not yet in the database.

Matikainen et al. propose an approach with the same data in entries for the second date (DSM and CIR Orthophotos), but with a raster version of the database (Matikainen et al. 2007). This method follows the same reasoning as the previous one but sets up a classification tree for the parameters used to calculate similarity scores.

Olsen et al. also use the raster version of the database for the second date as DSM and CIR orthophotos for the first one (Olsen and Knudsen 2005). Information on the evolution of buildings are obtained in a similar manner to the method proposed in Champion (2007) but further processing is carried out in order to keep only the modified buildings.

Finally, Rottensteiner present a method giving similar results but requiring a DSM as a minimum input for the second date and that can both process a raster and vector database (Rottensteiner 2008). A comparison of these methods is proposed by Champion et al. (2009). This article demonstrates that a method relying on images, even partially, induces to implement complex methods.

We propose a new method based on 3D data that are already extracted. The comparison will take place on vector data taking into account semantic information in CityGML and cadastre files; data images related to this area are no longer necessarily available.

Another approach, rather based on terrestrial image data, is proposed by Taneja et al. (2013). From 3D Cadastral data (which could be likened to our CityGML LOD2) and panoramic images from Google StreetView, this method gets building changes displayed on panoramic images. Although the results are still incipient, this already shows the interest of such 3D data, which are now democratized.

Sasagawa et al. (2013) propose another comparison between two dates, each of which needs a DSM and Orthophotos . After comparing the two DSM and Orthophotos, the results have been correlated to extract polygons for areas which have changed in elevation. However, the comparison between the DSM is still too sensitive to vegetation while the extracted polygons are not precise enough: they may include several neighbouring buildings. This last problem is a characteristic of comparisons based on extraction of buildings from automatically generated data. They do not differentiate two neighbouring buildings with geometric and visual characteristics which can be continuous. It is also to enhance this last point that we consider the contribution of the cadastral data important to improve the accuracy of the result of our comparisons.

A last way to detect changes is to take into account 3D geometry. Comparing 3D geometry is a well-known problem in geometric modeling. Methods based on the Hausdorff distance can be used to measure errors between two 3D models (Cignoni et al. 1995). Unfortunately, these methods cannot be used on the entire mesh of the scene because it is first necessary to decompose buildings blocks and to pair buildings from the two dates.

The methods studied above demonstrate that it is difficult to implement comparison methods based on the same kind of data. It is often necessary to aggregate several types of data to improve the results of change detection methods. Our approach is more pragmatic since we choose to use 3D data in CityGML files input and possibly to complete the process of buildings detection with cadastral information. This problem was born from the need to automatically detect changes in several versions of the 3D city made available by the city of Lyon.

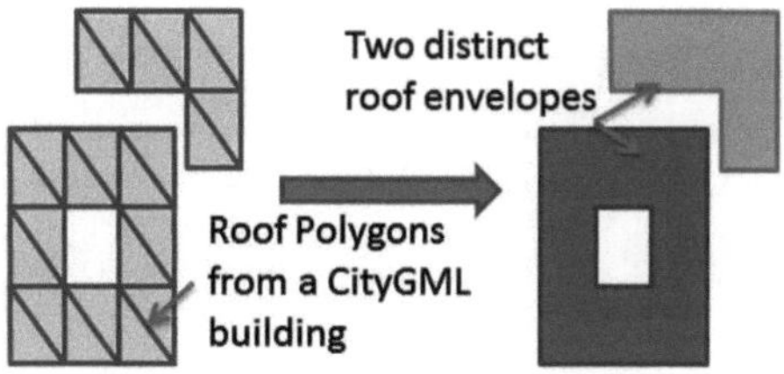


**Fig. 2** Creating polygons forming the footprint

## 3 Generate Building Footprints from CityGML Data

Our method is partially based on using building footprints. This operation is available in several software such as RhinoCity (Rhinocity 2014). However, we propose to present the method implemented in 3D-Use, which is a software developed in our team.

The CityGML files are composed of objects of the city and especially buildings. In general, walls and roofs polygons are described separately. The walls are actually calculated with information related to roofing. So we rely on the roofs, which are most directly taken from native data, to generate a 2D projection of the buildings and get their footprints. Note that this data may already be present in the CityGML and listed as "ground". This means that our first step is no longer necessary in this case.

The proposed method is simple; polygons belonging to the roof of buildings are projected on the Oxy plane (ground). This projection method is based on the principle that two polygons belonging to roofs cannot be superimposed. This is equivalent to assume that the wall only have polygons orthogonal to plane Oxy, which is true for the provided data. This is generally true as long as we remain in LOD2 but could cause problems in higher LODs.

For each building, a merging algorithm of projected roofs polygons is then used. So we get a bounding (envelope) polygon per disjoint element. Holes may appear in created envelopes. They come from courtyards belonging to groups of buildings and can be removed if necessary by our algorithm. An illustration is given in Fig. 2 with a simplified CityGML building composed of two parts cut into polygons. Roof was projected on the Oxy plane and it provides two sets of polygons (left). By merging neighbouring polygons we obtain two separate polygons (right).

## 4 Measuring the Development of a City with Two CityGML Files

CityGML data are usually generated periodically and, although acquisition frequencies are of the order of several years, it may be interesting to measure the evolution of a city and its buildings between two dates. In this purpose, we propose

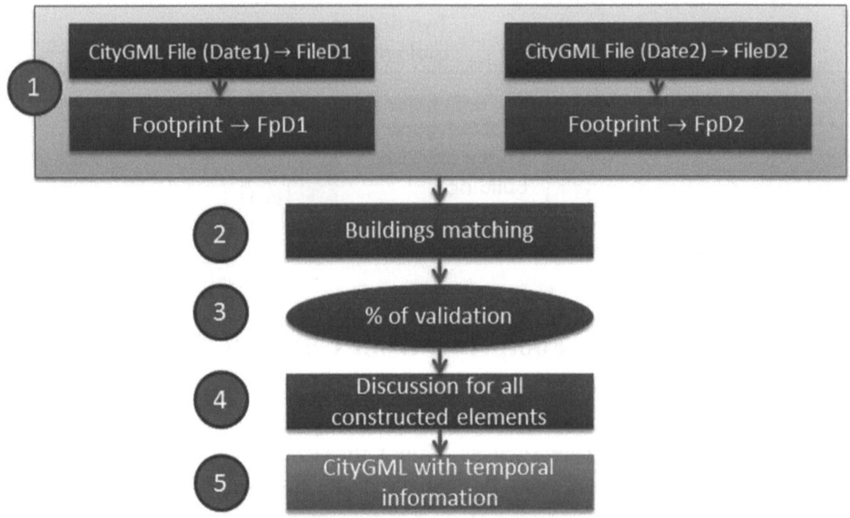


**Fig. 3** Steps to measure changes between two CityGML files

a method in this section based on the comparison of two CityGML files of a same area obtained at two different dates (FileD1 and FileD2). We assume that data from the two files have the same coordinate reference system. A conversion would thus be required if the two systems are different. The detection process of this evolution is described in Fig. 3.

The previous section described how the footprints are obtained from CityGML files. This part explains the following steps.

In step 2, we pair intersecting polygons in FpD1 and FpD2. If a match is found between two buildings $B_1$ and $B_2$, then the similarity must be quantified (step 3 and 4). Two tests are required. Let P be the intersection of $B_1$ and $B_2$. The first test involves comparing the area of the polygon P with the area of $B_1$ and $B_2$.

Let $Q1 = \frac{Area(P)}{Area(B_1)}$ *and* $Q2 = \frac{Area(P)}{Area(B_2)}$.$B_1$ and $B_2$ are considered potentially unchanged if $Q1 > 0.99$, $Q2 > 0.99$, $Area(B_1) - Area(P) < 5$ (m$^2$) and $Area(B_2) - Area(P) < 5$(m$^2$).

Otherwise $B_1$ and $B_2$ are both identified as “Modified Building” if $Q1 > 0.5$ and $Q2 > 0.5$. If those conditions are not verified, $B_1$ and $B_2$ will be respectively identified as “Destroyed Building” and “Constructed Building”.

We illustrate this test with Fig. 4, where we intersect blue and red buildings (respectively $B_1$ and $B_2$). This intersection gives three areas: A, B, C (in m$^2$). Our test will consider $B_1$ and $B_2$ as:

- Unchanged if: $A < 0.01 * C \&\& B < 0.01 * C \&\& A < 5 \&\& B < 5$(1)
- Modified if: (1) is false $\&\& A < 0.5 * C \&\& B < 0.5 * C$ (2)
- Destroyed/Constructed if: (1) and (2) are false (equivalent to $A \geq 0.5 * C$ *or* $B \geq 0.5 * C$)

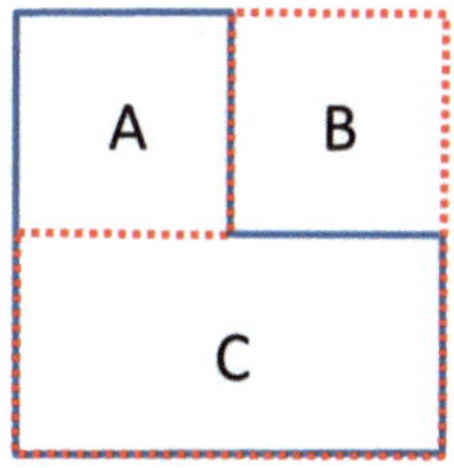


**Fig. 4** Intersection between *blue* and *red* buildings gives three areas: A, B and C (shared area)

We introduce another check in order to compare the linked 3D data and validate the first 2D result for buildings considered potentially unchanged. We have thus implemented a new test based on Hausdorff distance to measure the 3D distance between two buildings with the same footprint. This method takes into account all the 3D points composing roofs: for each point, we calculate a distance between it and the roof of the other building. The Hausdorff distance is the maximum value from all these calculated distance. It could express a change between the two buildings while it was non-detectable with a footprint comparison. This second test thus allows us to extract identical buildings that will be identified as "Unchanged building". Others will be added to the "Modified building" list.

To take into account the possible difference of accuracy between the two datasets, we should add a precision-dependent value to all our thresholds. This work is currently in progress.

Figure 5 shows an example of a building where the comparison of 3D data is needed. The first row shows the reference building and two different evolutions with a same footprint. On this illustration, the distance from a point to the reference building is represented in color ranging from blue (null distance) to red (maximum distance). 3D visualizations have been obtained with Cloud Compare (2014).

# 5 Improve the Change Detection Using Cadastre

## 5.1 *Overall Presentation of the Method*

Section 4 proposed a method of change detection on buildings present in CityGML files. Unfortunately, a number of buildings can actually refer to a set of neighboured buildings. In the initial stage of the footprints computation, groups of buildings have already been cut by separating the disjoint part (Fig. 6, left). Some sets are however still too important, inducing change detections while only a part of this set has actually been changed (Fig. 6, right). The top right image corresponds to 2009 data; bottom right is dedicated to 2012 data. It is necessary to propose an improvement of our method, based on the cutting of buildings using cadastral data.

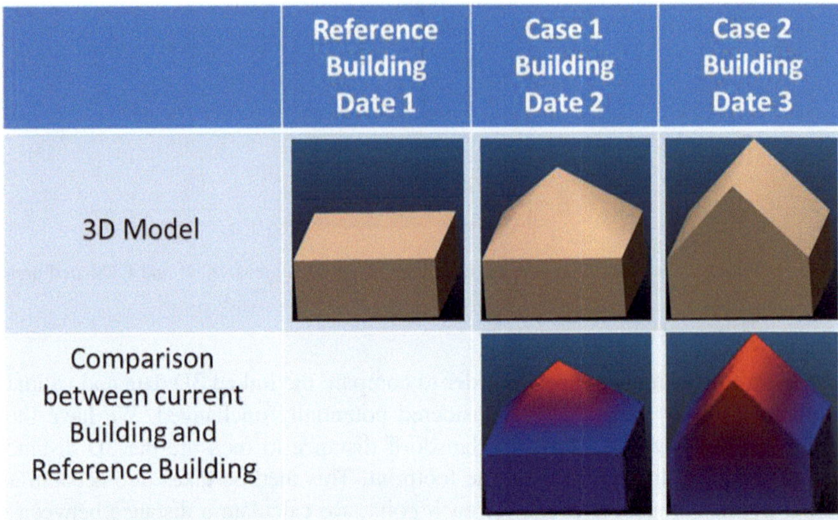


**Fig. 5** Calculation of the Hausdorff distance. The *top left* building serves as a reference. The two others represent two different dates of evolution. Results are proposed in the *second row*

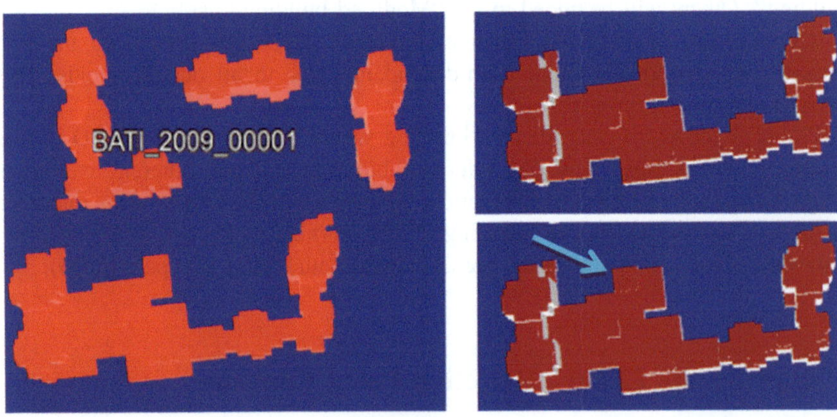


**Fig. 6** Set of buildings of significant size that is necessary to cut out. *Left* disconnected parts with a unique "Id". *Right* the comparison of the same set of building at two different dates shows that an additional part (highlighted with an *arrow*) has appeared between 2009 (*top*) and 2012 (*bottom*)

The method proposed is presented in Fig. 7. The first step is to bind buildings present in CityGML files with the cadastre in order to set up this cutting. Cadastral data are constituted of 2D polygons with various semantics information. We recall that this comparison is also motivated by the need of improving the semantics of CityGML data with information related to the cadastre. It may contain, for example, its date of construction or the name of its owner.

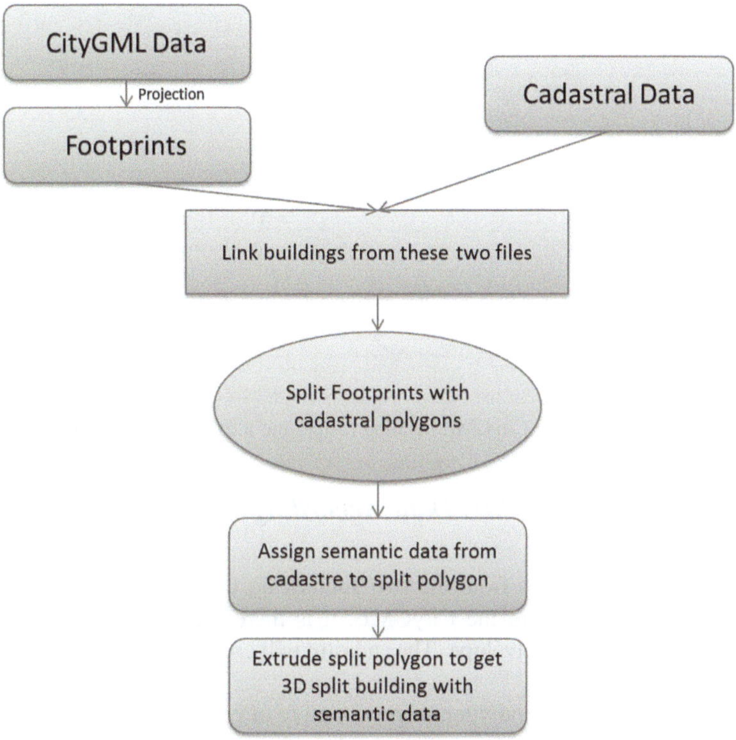


**Fig. 7** Matching of the data present in the CityGML data with the cadastre

Again, our matching method is based on building footprints. Let $P_1$ be the polygon corresponding to the footprint of one of the buildings in CityGML, $P_2$ a polygon from the cadastre paired with $P_1$, and $P_3$ the intersection between these two polygons. This intersection will be kept only if the following property holds $\frac{Area(P3)}{Area(P1)} > 0.1\, or\, \frac{Area(P3)}{Area(P2)} > 0.1$. This relationship is illustrated in Fig. 8. A footprint of a building from CityGML is presented on the left. Elements of the cadastre are superimposed on this building on the right.

Figure 9 shows an example of differences between a CityGML building (cyan) and a cadastral building (magenta) which represents same building.

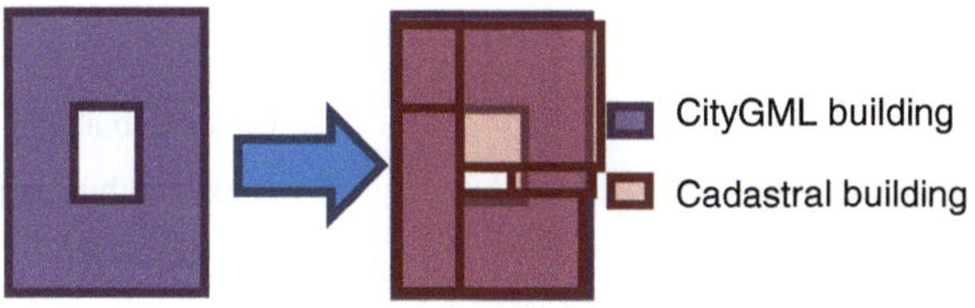


**Fig. 8** Connection between a CityGML building and those from the linked cadastre

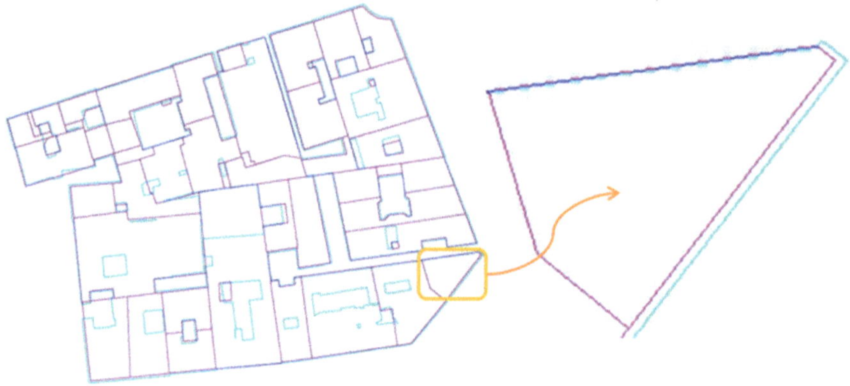

**Fig. 9** Difference between a CityGML building (*cyan*) and a cadastral one (*magenta*)

## *5.2 Cutting the CityGML Data According to the Cadastre*

We can see that polygons from the two sources do not match properly. Our aim here is to cut according to the CityGML; it is therefore necessary that cadastral polygons cover its outer envelope. The cadastre will create cutting lines between the buildings.

From this perspective, polygons representative of cadastral buildings are dilated to maximize the areal proportion of building from CityGML they cover. After removing overlapping areas that may appear between two cadastral polygons, we get the result shown in Fig. 10.

The next step is to get 3D data considering cuts deducted from the cadastre. The link between footprint and CityGML data has been kept. It thus becomes possible to deduct which roof element corresponds to each cadastral building. By applying this method to each polygon corresponding to a cut building, we extrude their footprints to create new 3D polygons consistent with the initial CityGML.

These new polygons are listed as roofs in the new CityGML and the missing walls are generated from them.

In the end, buildings which shapes are consistent with the cadastre are obtained. This information is added in the CityGML file (Fig. 3, step 5). It is then possible to use this element to compare CityGML files while the reduced size of the buildings

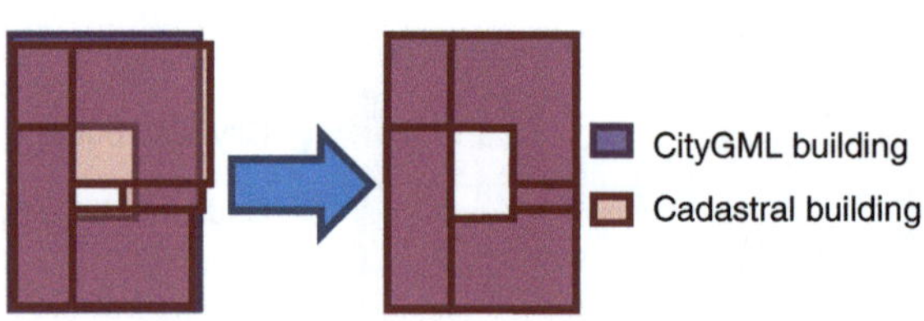


**Fig. 10** Projection of cadastral buildings on the linked CityGML polygon

allows us to generate more accurate results. For example, when a set of buildings was formerly listed as "Modified building", it is now possible to define which of them that actually were.

# 6 Results for Change Detection Between Two CityGML

We illustrate this method on two datasets. The first one concerns a former military camp where many areas have been modified (Sathonay Camp dataset). The second dataset is a part of the "campus Universitaire de la Doua" in Villeurbanne. On this second test set, a file containing the positions of modified buildings is at our disposal. Thus, it will be possible to validate our method with these data from a manual acquisition campaign.

Figure 11 gives an overview of the two CityGML files of Sathonay Camp created in 2009 (left) and 2012 (right). Important changes are noticeable in the middle of each image. The result of the comparison is presented in Fig. 12.

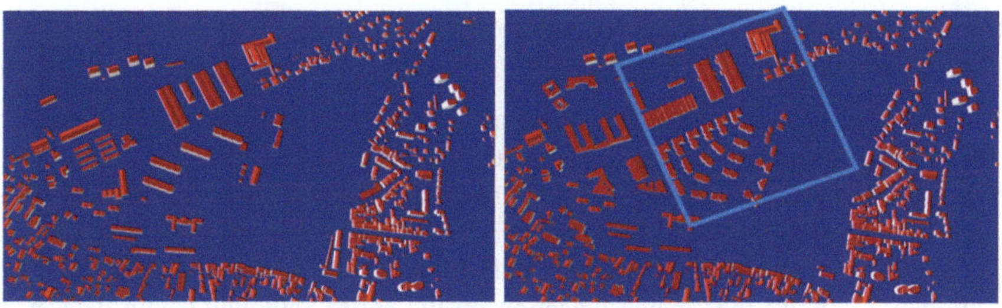

**Fig. 11** Sathonay camp CityGML. *Left* 2009 data. *Right* 2012 data

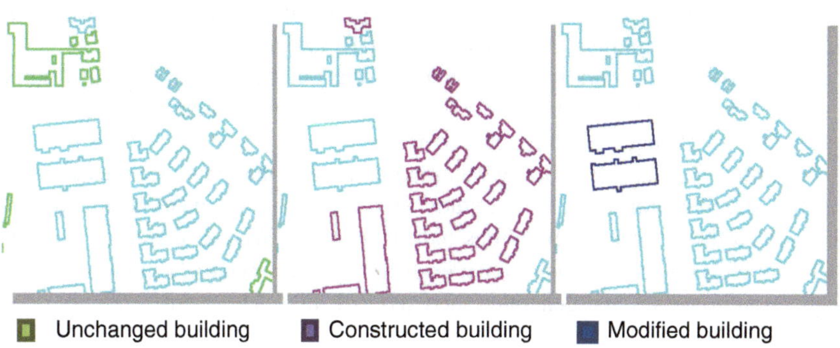


**Fig. 12** Comparison on sathonay camp (zoom in on the presented box Fig. 11, right). *Left* unchanged buildings are in *green*. *Middle* created buildings are in *purple*. *Right* changed buildings are in *dark blue*

**Fig. 13** 3D visualization of changes on sathonay camp

These results indicate that this area has greatly evolved. However, our method allows us to differentiate the buildings that were modified from the constructed ones. A visualization of the information obtained is possible using our software 3D-use (Fig. 13). Buildings that have been modified or created are represented in red and are superimposed on the textured data from 2009.

Figure 14 shows the change detection on the entire Sathonay Camp dataset. It describes the state of each building extracted from the two CityGML files.

Figure 15 gives the result of our method on the dataset "la Doua campus". Unchanged buildings are in green while the changes appear in cyan.

Figure 16 shows the locations of modified buildings manually identified and superimposed on building footprints of the 2012 dataset.

All identified areas correspond to buildings that our method has called "Modified building" or "Constructed building". This means that in this area, buildings will not be tagged "Unchanged building" by error.

Nevertheless, our algorithm reveals modified zones that were not present on the manual identification. After visual study of these areas, it appears that there were changes on these forgotten buildings but often minor. We can therefore wonder if these imprecisions come from human error or from an inaccuracy in the successive acquisition processes that generated the CityGML files. These errors are not due to our algorithm.

**Fig. 14** Change detection on sathonay camp between 2009 and 2012

**Fig. 15** Result of our method on the campus of la Doua in Villeurbanne

# 7 Adding Cadastral Information

We illustrate the functioning of our cutting method by looking at a building from a CityGML file of the city of Lyon. It is thus identified as a building with a single ID, while on the same area a large number of polygons are present in the cadastre (Fig. 17, left). The polygon of the building footprint from the initial CityGML is displayed in green while polygons corresponding to cadastral buildings are in magenta. We have applied our method to all these cadastral polygons in order to offer a consistent cutting of the original building (Fig. 17, right). This set of buildings obtained will then improve a possible comparison with data from another CityGML file. It can also simply be used to generate a more accurate version of this set of buildings, containing new semantic information, which is added to the CityGML file.

By extruding the information in 3D, we now have a precise cutting of the buildings. Thus, when we had a group of buildings identified as only one building by the CityGML (Fig. 18, left), we now have a set of separated buildings containing semantic information from the cadastre (Fig. 18, right).

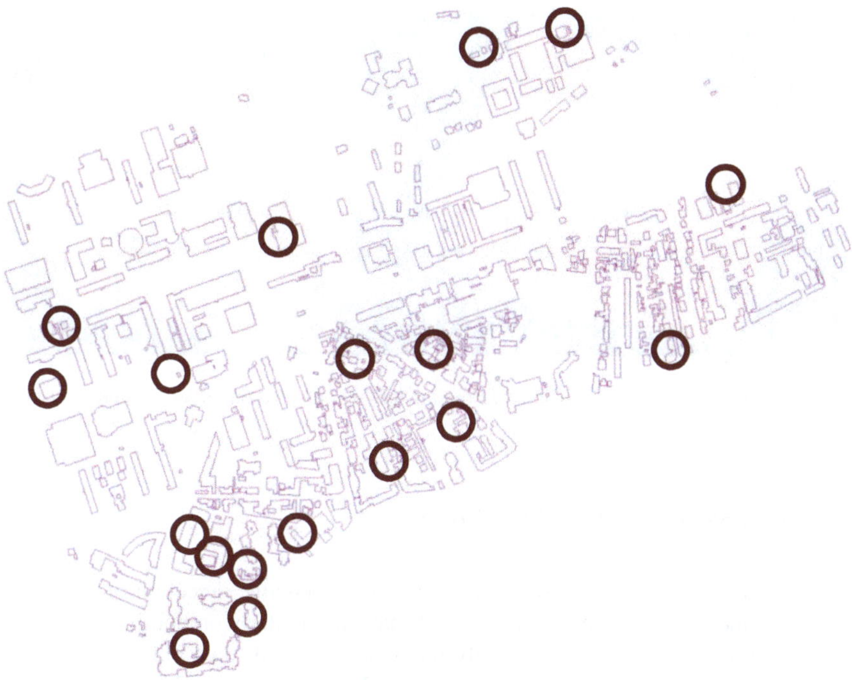

**Fig. 16** Location of the modified buildings manually identified

**Fig. 17** Cutting according to the cadastre. *Left* juxtaposition of the cadastre with the CityGML data. *Right* result cutting

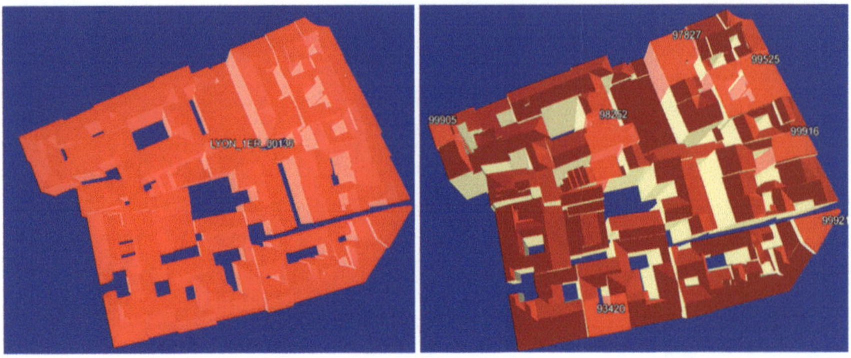

**Fig. 18** This unique building in CityGML (*on the left*) was divided into a set of buildings (*on the right*)

## 8 Conclusion and Future Works

In this paper, we have proposed a new method for change detection on urban 3D data. This method was applied to CityGML data, particularly on buildings, acquired in 2009 and 2012. The approach only relies on CityGML data unlike many other methods based on aerial or terrestrial images. This approach finds its full meaning in the lifecycle management of the 3D data for versioning or more generally in applications where temporal data are needed.

Our method gives very good validated results, as demonstrated in the Sect. 6. An improvement of the method was proposed to decompose buildings with the cadastre in order to possibly better isolate areas that have evolved over time. Beyond this use of cutting buildings according to the cadastre, this improvement also allows us to add semantic information to each piece of building cut. This method is based on the use of LOD2 building from CityGML files because our data are so constituted. However, only a few changes should be necessary (which would mainly affect footprints generation) to adapt it to LOD1 and LOD3 while LOD0 is equivalent to footprints we already compare.

The prospects of this work are numerous. As a first step, this method must be improved in order to make it fully automatic; it will be necessary to propose strategies of automatic calculation of the different thresholds. A second goal is to add topological information to enhance our method. For instance, the establishment of a topological structure as in Diakité et al. (2014) would allow us to obtain more precise information of the evolution of a building. In addition, it seems necessary to refine our cutting algorithms of CityGML data in order to directly store buildings obtained by comparison of CityGML data with the cadastre. In particular, a cut of textures has to be proposed. Finally, in the article proposed by Morel (2014), each state of an object is connected to a date. Upon detection, we can only conclude that a building appeared between a date and another. It is then necessary to relate the state of an object to a date interval.

**Acknowledgments** This work was performed within the BQI program of Université Lyon 1. This work was also supported by the LABEX IMU (ANR-10-LABX-0088) of Université de Lyon, within the program "Investissements d'Avenir" (ANR-11-IDEX-0007) operated by the French National Research Agency (ANR). CityGML data are provided by "Le Grand Lyon". Special thanks to Hugo Sahuquet who has carefully read this paper.

# References

Champion N, Rottensteinerb F, Matikainenc L, Liangc X, Hyyppäc J, Olsend BP (2009) A test of automatic building change detection approaches. In: CMRT09: object extraction for 3D city models, road databases and traffic monitoring—concepts, algorithms, and evaluation

Champion N (2007) 2D building change detection from high resolution aerial images and correlation digital surface models. In: PIA07—photogrammetric image analysis

Cignoni P, Rocchini C, Scopigno R (1995) Metro: measuring error on simplified surfaces. Comput Graph Forum 17(2):167–174

Cloud Compare (2014) http://www.danielgm.net/cc/

Diakité AA, Damiand G, Van Maercke D (2014) Topological reconstruction of complex 3D buildings and automatic extraction of levels of detail. In: Eurographics workshop on urban data modelling and visualisation

Matikainen L, Kaartinen K, Hyyppä J (2007) Classification tree based building detection from laser scanner and aerial image data. In: ISPRS workshop on laser scanning 2007 and silvilaser 2007

Morel M, Gesquière G (2014) Managing temporal change of cities with CityGML. In: Eurographics workshop on urban data modelling and visualisation, pp 37–42. doi:10.2312/udmv.20141076, ISBN:978-3-905674-49-1

Olsen B, Knudsen T (2005) Automated change detection for validation and update of geodata. Poster session presented at cognitive vision summer school Bonn, Bonn, Aug 2005

Rhinocity (2014) http://www.rhinoterrain.com/en/rhinocity1.html

Rottensteiner F (2008) Automated updating of building data bases from digital surface models and multi-spectral images. In: IAPRSIS XXXVII—B3A, pp 265–270

Sasagawa A, Baltsavias E, Aksakal SK, Wegner JD (2013) Investigation on automatic change detection using pixels-changes and DSM-changes with ALOS-PRISM triplet images. In: International archives of the photogrammetry, remote sensing and spatial information sciences, vol XL-7/W2

Taneja A, Ballan L, Pollefeys M (2013) City-scale change detection in cadastral 3D models using images. In: Computer vision and pattern recognition (CVPR), 2013

# Advances in Structural Monitoring by an Integrated Analysis of Sensor Measurements and 3D Building Model

**Thomas Becker, Sven Weisbrich, Cheng-Chieh Wu and Frank Neitzel**

**Abstract** The use of open GIS standards offers a broad variety of potential, particularly in the field of data exchange, data storage, and interoperability. GML and CityGML are excellent examples for the ontological description of real world objects by means of an open standard whereas SensorML serves to describe measurements, sensors and measuring platforms. The use of such standards offers not only the possibility of using a common standardised language, but also the use of open service standards. The combination of spatial data and sensor standards in services and service-oriented architectures goes far beyond previous existing solutions on the market and provides a novel platform for monitoring structures. That in fact is far more than a simple data storage model. The methods and models presented in this contribution allow a direct integration of sensor data and its provision through an open standard language. In this case, all the intermediate steps at any time through an open service interface are addressed and may be made available and provided to different actors and stakeholders participating in a construction scenario. The great potential and the added value of such an information system is the permanent availability of measurement and object data and an associated integrated analysis of sensor data in combination with a finite element model (FEM). The automatic derivation of a finite element model from the 3D structure model, the visualisation of FEM, the provision of raw (measurement) data and sensor information for each time of measurement transform the platform into a

T. Becker (✉) · S. Weisbrich · F. Neitzel
Institute of Geodesy and Geoinformation Science, Technische Universität Berlin, Berlin, Germany
e-mail: thomas.becker@tu-berlin.de

S. Weisbrich
e-mail: s.weisbrich@tu-berlin.de

F. Neitzel
e-mail: frank.neitzel@tu-berlin.de

C.-C. Wu
Division 8.1—Sensors, Measurement and Testing Methods, Federal Institute for Materials Research and Testing, Berlin, Germany
e-mail: cheng-chieh.wu@bam.de

© Springer International Publishing Switzerland 2015

M. Breunig et al. (eds.), *3D Geoinformation Science*,
Lecture Notes in Geoinformation and Cartography,
DOI 10.1007/978-3-319-12181-9_9

universal tool in the field of structural monitoring. This contribution introduces the individual components, the standards used and the interaction between the components to an overall system.

**Keywords** Structural monitoring · SensorML · Finite Element Method · Integrated analysis · City model · Building model

## 1 Motivation

A main application in the field of geodesy and one of the most important tasks in engineering geodesy is the monitoring of structures. Since at the beginning site visits to pick up the measured data was mandatory. Nowadays open standards, the internet-of-things (IoT) and the OGC Sensor Web Enablement (SWE, Bermudez 2011) allow the development of web based solutions. A permanent monitoring covers the measurement of changes of various parameters such as, deflection, inclination, strain, temperature, humidity etc. The measured parameters are both at points distributed over the entire structure, as well as holistic, considering the complete structure. Especially after or even during extreme events structures can be controlled in real time by using already installed sensor networks (Boller and Staszewski 2004; Botts et al. 2006; Resch 2012; Bröring et al. 2011, 2012).

Material degradation or the change of geometric properties such as boundary conditions and system partial connectivity can largely affect the structural performance and safety. Maintaining service will be impossible without knowing and analysing the potential damage. Today structural condition assessment is the prerequisite for prediction of engineering structures remaining in service (Farrar and Worden 2007; Worden and Barton 2004). The initial step for monitoring is the periodically or permanent observation of the behaviour of a structure under external forces, e.g. wind, earthquakes, traffic etc., by measurements of heterogeneous sensors (Welsch et al. 2000, p. 45). The main important part is to analyse these measurements under consideration of the mechanical model of the structure via least-squares adjustment (Jäger 1988; Neitzel et al. 2014; Lienhart 2007; Teskey 1988). This allows further analysis of structural properties by means of statistical tests and thus a reliable assessment of its condition. The capability of real or near real time sensor data access can largely affect the performance of the information system for structural monitoring. However, the automatic process of the high sample rate sensor data is still an ongoing development for most products on the market.

Permanent monitoring systems vary in size, shape, and sensors substantially but have almost common requirements and features. Open access to the measurements and data management as well as its analysis and interpretation are key issues in a geodetic monitoring.

The use of open standards, particularly in the area of web services facilitates an excellent opportunity to provide all involved parties a uniform, standardised machine-readable format. Using the Sensor Web Enablement (SWE) standards of the Open Geospatial Consortium (OGC), all kinds of sensors, transducers and sensor data repositories can be located, accessed and used via the WWW. The combination of sensor technology, computer technology and network technology, thus offers new solutions in the field of construction site safety, industrial controls, meteorology, geophysical survey, flood monitoring, risk assessment, tracking, environmental monitoring, defence, logistics and many other applications.

A system for structural monitoring must also offer the possibility for a visual interpretation of the measured data as well as the analysis results. Besides a treatment in tables, graphs, or text form is sufficient for the representation of the measured data, the presentation of the analysis results requires the inclusion of the structure itself.

The presented approach will demonstrate how open standards such as CityGML (Gröger et al. 2012) and SensorML have to be combined in order to satisfy asset owner and asset manager needs. The proposed solution covers:

- Near real time measurement and storage of sensor data
- On-the fly analysis of measurement data
- Data exchange with stakeholders in a standardised way
- Conversion of CityGML to FEM and vice versa
- Integrated analysis of sensor data and FEM model based on CityGML standard
- CityGML extension proposal by material and time entities

It will be shown how two different standards can be harmonised and combined to fulfil an ongoing research task in the field of geodesy by applying methods and standards from the field of geoinformation science.

## 2 Requirements

Construction companies are demanded to guarantee for the operational capability of the structures they built. The use of sensor networks allows to monitor those structures in real-time. Without the use of sensor networks, a continuous monitoring is difficult to carry out and quite time and cost intensive. In addition to planning and performing geodetic measurements, the complete documentation, provision of data, analysis and results, captures a central position in the field of structural health monitoring (Furtner et al. 2013). Often, data have to be converted, reorganised, or specially prepared to meet all the requirements of stakeholders participating in a construction project. Thus, a structural health monitoring system will help asset owners and asset managers to minimise maintenance and observation costs by maximising the assets lifetime. An advanced tool for structural health monitoring based on an integrated analysis has to cover the following requirements:

1. **Data storage**
   The huge amount of measured sensor data has to be structured and stored in a redundant database. Moreover, the database and related interfaces should enable easy access to the stored data via web services, such as OGC's SOS (Bröring et al. 2012).
2. **Data pre-processing and preparation**
   The system should provide an analysis tool in order to detect malfunctions of sensors or to identify gross errors in the data. Furthermore, a detailed time series analysis of the sensor data should be provided, e.g. auto-correlation, cross correlation, detrending, filtering etc.
3. **Data and feature visualisation**
   An advanced system for structural health monitoring should also provide the possibility for a visual interpretation of the measured data and analysis results. Since a treatment in tables, diagrams or text form is sufficient for the representation of the measured data the visual presentation of the analysis results, requires the inclusion of the structure itself.
4. **Data aggregation and reduction**
   Asset owners and managers are mostly not interested in the huge amount of measured data. Thus, valuable filter operations, event detection algorithms, aggregation, and reduction methods are required to limit the transferred data to a feasible minimum.
5. **Integrated analysis**
   The integrated analysis via least-squares adjustment is based on a FE model that allows further analysis of structural properties by means of statistical tests and thus a reliable assessment of the current condition of the structure. The FE model is derived from existing 3D semantic object models, such as CityGML and the conversion from CityGML to FEM and vice versa is a major requirement of such a tool.

## 3 Integrated Analysis

To model the mechanical behaviour of a structure like a bridge, is in fact a quite challenging task. Besides its geometry many additional properties need to be taken into account, e.g. types of bearings, different materials, internal and external forces, temperature, etc. Therefore, the basic principle of an integrated analysis is demonstrated on a statically bended beam, as a simplification of the deformable deck of a bridge.

### 3.1 Mechanical Modelling

The deformational behaviour of statically bended beam under external forces can be described by the specialised mechanical model of the Euler-Bernoulli beam theory. This theory yields a governing equation that connects all related and influential quantities together. The governing equation is mathematically expressed as an ordinary differential equation. In order to use the specialised mechanical model, we have to ensure that the tested beam specimen fulfils the following requirements: (a) In the bent configuration, the cross sections of the beam remain plane and perpendicular to the neutral axis—the set of lines that don't extend or contract. (b) The beam is only subjected to an external pure bending moment—shear forces, axial normal forces or torques are excluded. (c) The beam consists of isotropic material which obeys Hooke's law. (d) The total length of the beam is much greater than the deflection. (e) The dead load of the beam can be neglected. Under above requirements the governing equation reads as

$$\frac{\mathrm{d}^2 u(x)}{\mathrm{d}x^2} = -\frac{M(x)}{EI}, \tag{1}$$

where $u$ is the deflection of the beam at a position $x$, $M$ the bending moment, $E$ the Young's modulus and $I$ the moment of inertia. The differential equation (1) describes the deformational behaviour of the beam under a bending moment and can be solved analytically. Oftentimes a mathematical exact solution for more complex structures is not found yet or very time-consuming to compute by hand. In this case, numerical methods such as finite element method are introduced (Zienkiewicz 1971). They transform partial differential equations into linear algebraic systems that can be solved with computers with known algorithms.

For a numerical example we generated a chain-like FE model with 726 elements of a four-point bending test which is illustrated in Fig. 1. The properties of the aluminium beam specimen and the forces the specimen is exposed to are listed in Table 1.

In order to include inclination and strain for each element, we approximate its deflection by a polynomial function of 5th order. The approximated solution of the differential equation (1) yields

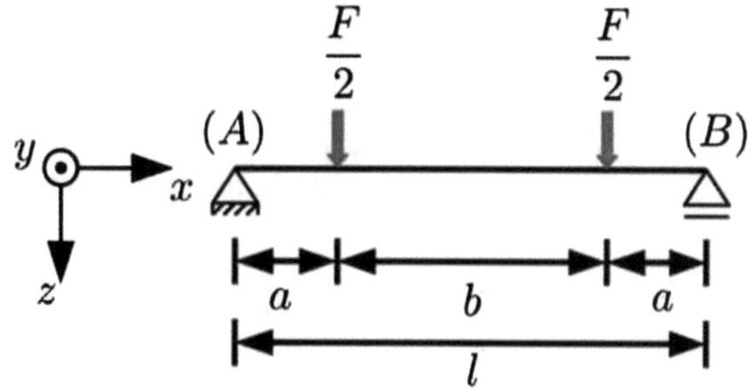


**Fig. 1** Four-point bending test

**Table 1** Properties of the beam specimen

| | |
|---|---|
| Length $l$ | 7.26 m |
| Width | 0.20 m |
| Height | 0.36 m |
| Elastic modulus | 70 GPa |
| Force $F$ | 7,460 N |
| Force position $a$ | 2.42 m |

$$\mathbf{u} = \mathbf{K}^{-1}\mathbf{f}. \tag{2}$$

The matrix **K** is known as stiffness matrix and the vector **f** is denoted as load vector. The vector **u** contains the displacement, inclination and strain at each node of the FE-model. The **K** matrix only depends on the number of elements of the FE-model, while the load vector **f** is a function of the parameters of the experimental setup. Therefore, the behaviour of the beam can be determined. Such a kind of problem definition is known as a forward problem.

## *3.2 Data Analysis*

The described forward problem usually occurs during the design phase, while the engineer determines the behaviour of a structure under external forces. However, during a monitoring usually the behaviour of a structure is observed at different positions using different types of sensors. The analysis of those hybrid measurements is directly based on the solution (2) of the differential equation (1) and reads as

$$\begin{bmatrix} u_0 \\ \vdots \\ u_i \\ \vdots \\ u_N \end{bmatrix} = \begin{bmatrix} k_{00} & \cdots & k_{0j} & \cdots & k_{0N} \\ \vdots & \ddots & \vdots & \ddots & \vdots \\ k_{i0} & \cdots & k_{ij} & \cdots & k_{iN} \\ \vdots & \ddots & \vdots & \ddots & \vdots \\ k_{N0} & \cdots & k_{Nj} & \cdots & k_{NN} \end{bmatrix} \begin{bmatrix} f_0 \\ \vdots \\ f_j \\ \vdots \\ f_N \end{bmatrix}. \tag{3}$$

The functional model for a measurement $u$ at a node position $i$ yields

$$u_i = k_{i0}f_0 + k_{i1}f_1 + \cdots + k_{iN}f_N, \tag{4}$$

in which the load vector $f_j$ contains the elastic modulus $E$

$$f_j = b_j \frac{1}{E}. \tag{5}$$

While choosing the elastic modulus $E$ as an unknown parameter results in the functional relationship

$$u_i = \left( \sum_{j=0}^{N} k_{ij} b_j \right) X \tag{6}$$

between the measurement $u$ at a node position $i$ and the unknown parameter $X = 1/E$ and is known as inverse problem. The values $k_{ij}$ and $b_j$ are treated as error free. Furthermore, the stochastic model needs to be set up in order to take the individual stochastic properties of the different types of observations into account. Introducing more observations than unknown parameters yields an over determined equation system which can be solved via least-squares adjustment and thus allows a statistically analysis of the observations and unknown parameters.

## 3.3 Sensor Placement

To develop a sensor configuration best suited for the presented structure, we generated a finite element model with known boundary conditions and known constitutive law. Based on this model, we computed synthetic displacement, inclination and strain measurements for predefined measuring points. A Monte Carlo simulation (MCS) was performed to analyse the dependence of estimated parameters on the location of the measuring point as well as on the stochastic properties of the measurements. The results are depicted in Fig. 2.

As it can be seen, very accurate sensors are needed to allow a reliable estimation of the material property. Furthermore, it is obvious that displacement and strain sensors should not be located in the vicinity of the bearings, because in this area the measured value is much smaller than the assumed standard deviation of the sensors. For the position of inclination sensors the opposite is the case. The elastic modulus can be estimated best while measuring at the bearings. In addition, it can be shown that using more than 10 sensors hardly improves the accuracy of the estimated material property.

## 3.4 Damage Detection and Localisation

Furthermore the potential of this integrated analysis for damage detection and localisation within a slender beam due to local material degradation is presented. To simulate beam damage, we introduced for each element a separate elastic modulus. The extent of damage can be induced by a stepwise change of the elastic modulus for a predefined number of elements from 70 (undamaged) to 15 GPa (highly damaged). Starting from the 80th element of our chain-like FE model we increased

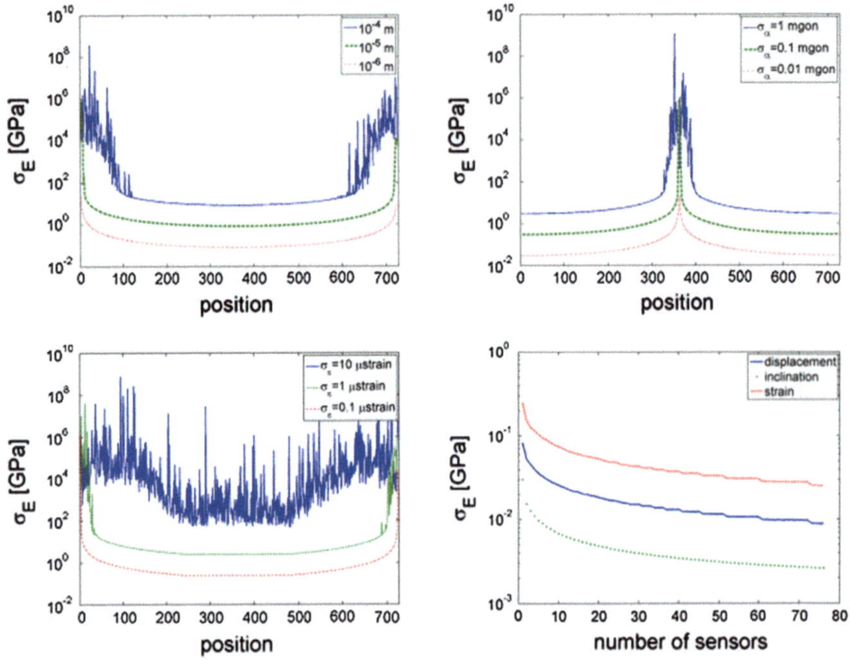


**Fig. 2** Accuracy of the estimated elastic modulus $\sigma_E$ depending on sensor position and accuracy of displacement (*top left*), inclination (*top right*) and strain (*bottom left*), and number of sensors (*bottom right*)

the number of damaged elements successively till the 89th element. For each simulated extent of damage, we calculated the displacement, inclination and strain of each node of the FEM. The calculated displacement and strain at node positions 70, 106 and 141 and inclination at 10 and 210 along the x-axis were used as true values for the synthetic measurements. Therefore a MCS of 1,000 experiments was performed. For each experiment we added normal distributed noise with a standard deviation of $\sigma_u = 1\,\mu\text{m}$, $\sigma_{u'} = 0.1\,\text{mgon}$ and $\sigma_{u''} = 0.5\,\mu\text{strain}$ to the true values. Based on those synthetic measurements and the functional model (6) we determined the elastic modulus for each element via least-squares adjustment in order to detect and localise the simulated damage. The results are shown in Figs. 3 and 4.

Figure 3 shows the potential of the new approach for an integrated analysis of spatially distributed hybrid measurements to detect the simulated damage. According to the chosen standard deviation of the synthetic measurement this approach is able to detect a change of the elastic modulus down to 40 GPa of two elements with a probability of 100 %. Furthermore the probability to localise this detected damage within a range of ±5 elements is ca. 55 %, as it can be seen in Fig. 4.

The presented approach for an integrated analysis based on a mechanical model of a structure is a fundamental component of an information system for structural monitoring.

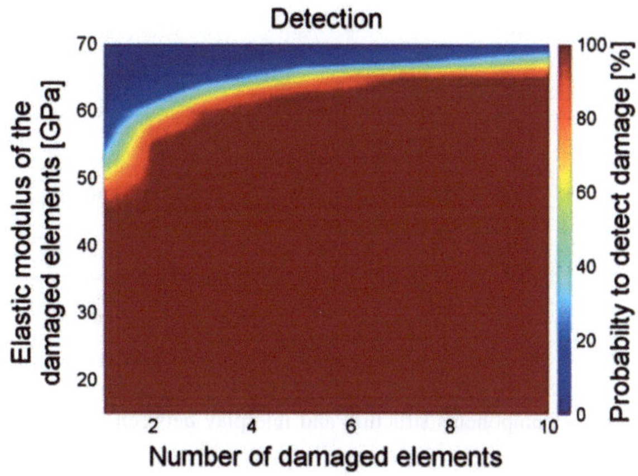


**Fig. 3** Damage detection depending on material degradation (change of elastic modulus) and number of damaged elements

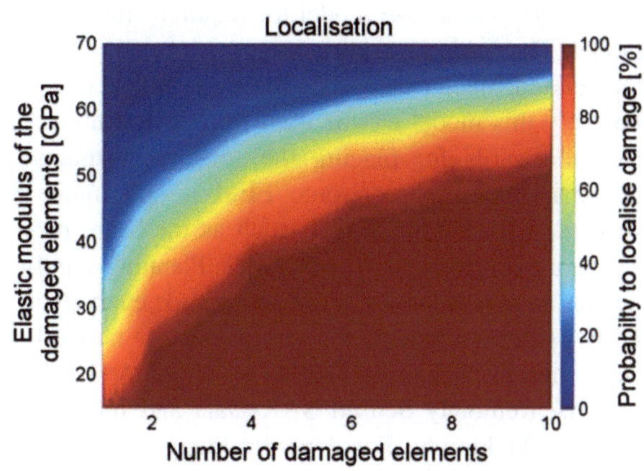


**Fig. 4** Damage localisation depending on material degradation (change of elastic modulus) and number of damaged elements

# 4 Systems Architecture

The architecture of this system has to be designed to enable the integration of individual components in such a way that the exchange of data between all components, but also the exchange with external partners via standardised interfaces is realised. That is a crucial point with respect to the upcoming realised INSPIRE directive (2007). The exchange of spatial data will play a big role in near future. As depicted in Fig. 5 the components mainly interact with each other via services and service requests.

This allows a very flexible use of the system either as desktop version in an office or as field system on a tablet computer. Moreover, the consequent usage of OGC SWE framework for sharing observation data and WFS for sharing spatial feature data enables participants and actors involved in a construction to retrieve needed data in a standardised way. As already mentioned in Chap. Modeling and

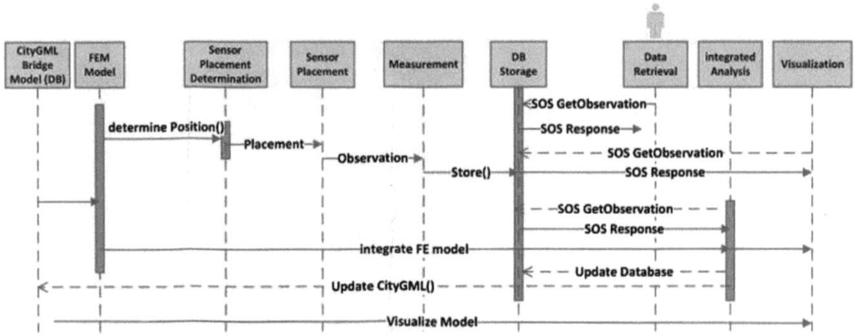


**Fig. 5** Components structure and interplay between the required components

Managing Topology for 3-D Track Planning Applications the CityGML bridge model serves as a base model for required site data and for the derivation of the FE model. The bridge model is stored in the 3D CityDB either in a Oracle database (Nagel and Stadler 2008; Stadler et al. 2008) or in a PostGIS DB (Kunde 2013). Instantly this model can be used for the determination of suitable sensor positions and afterwards for placing the sensors in reality. The mapping of CityGML boundary representation onto FEM is still under development. Since we were successful for simple objects such as the beam presented in Chap. Modeling and Managing Topology for 3-D Track Planning Applications a generic tetrahedronization approach has to be developed for every type of bridge supported by CityGML. After setting up the overall sensor network the observations are directly stored into the systems database, see Chap. A Hybrid Approach Integrating 3D City Models, Remotely Sensed SAR Data and Interval-Valued Fuzzy Soft Set Based Decision Making for Post Disaster Mapping of Urban Areas. Via an SOS interface, the data can be requested and responded in a standardised way such that every following component works in a service-oriented way (Erl 2008).

As shown in Fig. 6, the measured data of the sensors are directly stored in a GIS database—based on PostGIS or by using the 52° north framework—that works also on a PostGIS or Oracle database.

Both SOS encodings explicitly enable the integration of sensor observations using CSV or XML document in the database and thus provide the measurement data to the actors in an OGC-compliant exchange format. In terms of documentation, the raw measurement data are already available for each involved actor and can be used for tests and analysis—directly on the fly. Thus, both the sensor data and the observations can be provided in a standardised way and can be prepared for an integrated analysis based on finite element models (FENICS 2014). The results of such analysis can be directly represented in 3D and by means of an update function reintegrated into the database.

The management tool will be completely developed as a browser based service. The user interface can be seen as a front end, providing access to the observed data,

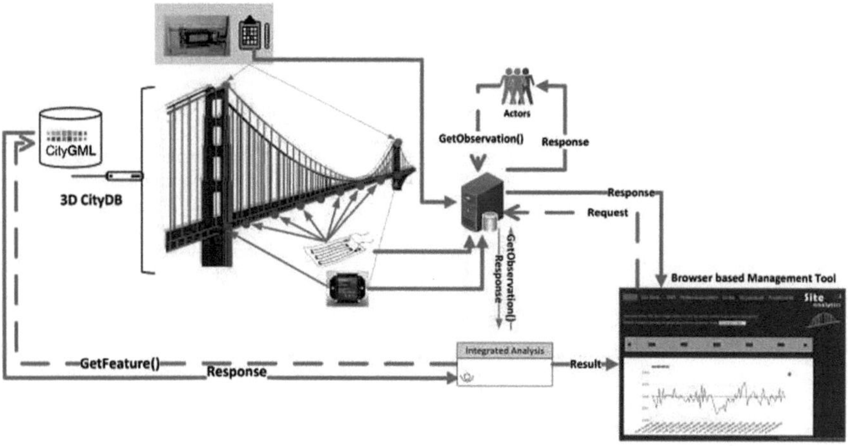


**Fig. 6** Systems architecture (part of the figure are under copyright of © microstrain.com, © 3dcitydb.org)

access to analysis tools and 3D and 2D data representation. This includes a general view on the observed data as kind of a dashboard supporting different types of chart diagrams and textual representation as well as the execution of statistical analysis tools for rapid data evaluation and long-term comparison of data.

# 5 Extending CityGML for Use in FEM

In order to derive a numerical solution computed from FEM simulations, three main input information of the target structure have to be provided from CityGML: Initial geometry, boundary conditions and material information.

The initial geometry is used to generate finite elements by subdividing the whole structure into parts with simpler shape such as cubes, tetrahedron etc. External influences which the structure is subjected to, are described by boundary conditions, for example dead load, bearing fixations, heat transfer, loads from vehicle, wind load and more. Furthermore, the materials the structure consists of need to be known. Since that information is not part of a typical CityGML model we have to extend CityGML by using the ADE mechanism in order to support FEM analysis. Currently CityGML provides three modules that are relevant for our kind of research—the buildings module, bridge module, and tunnel module. Since material information are needed for many applications besides structural health monitoring, it should be modelled as an own module, see Fig. 7. Thus, each *_CityObject* will have different kind of Materials. As first each *_CityObject* has an *ExteriorMaterial* describing the material type of the outer shell of an urban entity. Moreover a

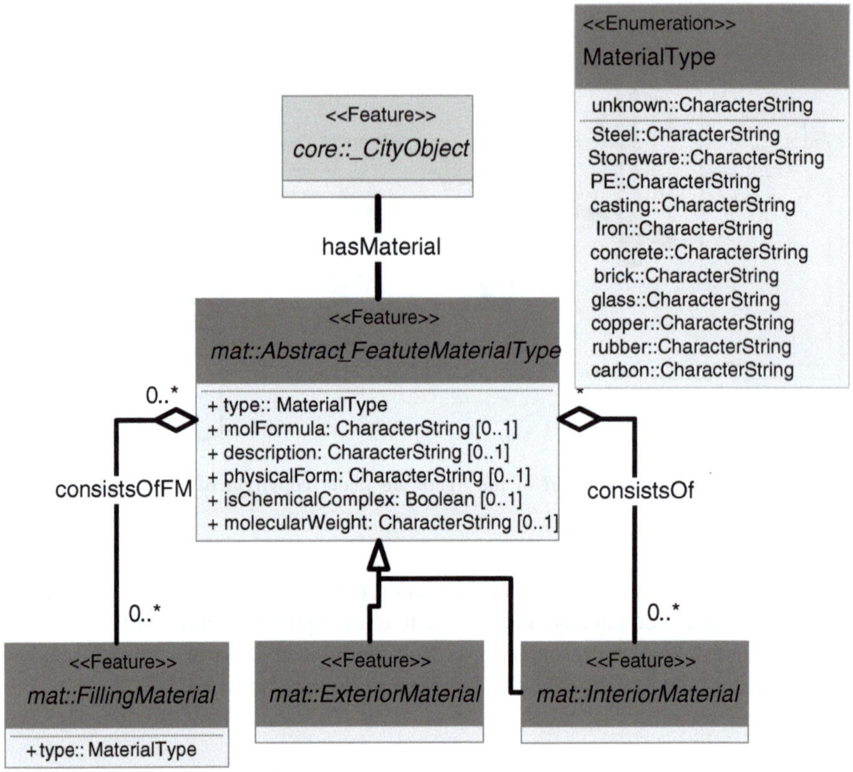


**Fig. 7** Proposal for integrating material information into CityGML

**Fig. 8** Material example to provide needed information for FEM

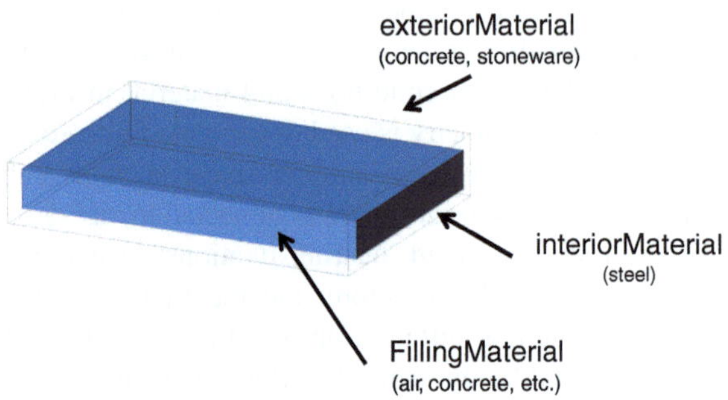


separation into *InteriorMaterial* and *FillingMaterial* is undertaken in order to describe the material of the inner shell of an entity as well as the filling material of the object itself see Fig. 8. The presented UML diagram (Fig. 7) just shows an initial approach, how to integrate material information into CityGML. Since various application nowadays also need such information for performing analysis and simulation e.g. energy and carbon dioxide calculations, bomb scenarios etc. this

**Table 2** Overview about relevant DBMS for designing the overall system

| Database name | MySQL | Oracle | Microsoft SQL Server Express | postgreSQL |
|---|---|---|---|---|
| Description | Widely used open source RDBMS | Widely used RDBMS | MS RDBMS | Based on the object RDBMS Postgres |
| Developer | Oracle | Oracle | Microsoft | PostgreSQL Global Development Group |
| License | commercial | commercial | Open Source | Open Source |
| Implementation language | C and C++ | C and C++ | C++ | C |
| Server operating systems | Linux, OS X, FreeBSD, Solaris, Windows | AIX, Linux, OS X, HP-UX, Solaris, Windows, z/OS | Windows | Linux, OS X, HP-UX, Solaris, Unix, Windows |
| DB model | Relational DBMS | Relational DBMS | Relational DBMS | Relational DBMS |
| Supported programming | Ada, C, C#, C++, D, Eiffel, Erlang, Haskell, Java, Objective-C, OCaml, Perl, PHP, Python, Ruby, Scheme, Tcl | C, C#, C++, Clojure, Cobol, Eiffel, Erlang, Fortran, Groovy, Haskell, Java, JavaScript, Lisp, Objective C, OCaml, Perl, PHP, Python, R, Ruby, Scala, Tcl, Visual Basic | .Net, Java, PHP, Python, Ruby, Visual Basic | .Net, C, C++, Java, Perl, Python |
| Server-side scripts | yes | PL/SQL | Transact-SQL and.NET languages | User defined functions |

will affect every module of CityGML. Thus more discussions and investigations have to be done until the integration of material information into CityGML can be stated as almost finished. This will be future work, especially for CityGML 3.0 where a special working group was formed to integrate material and texture information into the future version of CityGML.

## 6 DBMS and Performance Tests

The data generated by sensors are enormous and need to be stored very quickly into the database. Since enormous quantities of data is constantly being generated from sensors, […] "the spatial functionality offered by current commercial and open-source relational databases differs significantly in terms of available features, true geodetic support, spatial functions and indexing" (Ray et al. 2011). Thus, at first a revision of existing DBMS (compare Table 2) and the Pros and Cons should be taken into account followed by a performance test to ensure the thigh transaction speed. We decided to use PostgreSQL/PostGIS, since it is OpenSource, free available and widely used in GIS community.

However, the use of XML based databases is still a valid option. This is relevant for the overall decision of the design of the system, whether it can be a real-time system or not. A first rough test with a PostGIS database installed on a windows machine, using default setup parameters reveals that the DBMS is capable of handling nearly 6,000 entries per second by using 10–20 parallel working threads (cf. Fig. 9). Tuning the DBMS and allowing more parallel transactions, followed by giving more working memory, etc. should increase the number of entries per second dramatically. However, the current configuration is capable of handling a lot of low

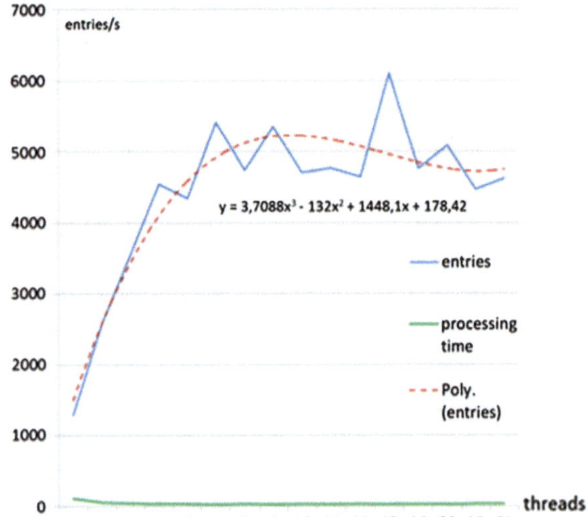


**Fig. 9** Initial performance test using a standard PostGIS installation in a windows machine

frequency (max. 10 Hz) sensors, such as strain, temperature, wind speed, etc. Applying high frequency sensors such as accelerometers will dramatically reduce the number of applicable sensors due to a measurement frequency of 512 Hz.

## 7 Conclusions

In this contribution, new possibilities for structural monitoring, analysis and simulation are presented that are significantly based on the interoperability of systems/ components. The combination of spatial data and sensor data/standards in services and service-oriented architecture provides a novel versatile platform for structural monitoring, which goes far beyond a simple data storage model. The method presented in this contribution allows direct integration of sensor data and its provision through an open standard language. The great potential of such an information system lies especially in the integrated analysis of the sensor data on the base of a finite element model (FEM). The automatic derivation of a finite element model from 3D construction model is until now only implemented for basic rectangular shapes. For complex structures new algorithms and methods needs to be developed. The visualisation of FEM—simulation results based on the construction model, the provision of raw data and sensor information for each point in time drives the platform into a universal tool in the field of structural monitoring.

## References


Bermudez L (2011) OGC sensor web enablement (SWE). https://www.evernote.com/shard/s264/sh/06758c15-b4cc-44fc-8440-2d26bf858604/e9e84a61e167e53bd1bc76f339934cc2/res/996f9a2d-4b7d-4670-9b40-3ad1ff6013ae/2011_OGC_Sensor_Web_Enablement.pdf. Accessed 22 April 2014

Botts M, Percivall G, Reed C, Davidson J (2006) OGC® sensor web enablement: overview and high level architecture. In: Nittel S, Labrinidis A, Stefanidis A (Hrsg) GeoSensor networks. Springer, pp 175–190

Boller C, Staszewski WJ (2004) Structural health monitoring. In: Proceedings of the second European workshop on structural health monitoring, München, pp 7–9

Bröring A, Echterhoff J, Jirka S, Simonis I, Everding T, Stasch C, Liang S, Lemmens R (2011) New generation sensor web enablement. Sensors 11(3):2652–2699

Bröring A, Stasch C, Echterhoff J (2012) OGC® sensor observation service interface standard, OpenGIS® Implementation Standard, version 2.0. http://www.opengis.net/doc/IS/SOS/2.0. Accessed 17 June 2014

Erl T (2008) Soa: principles of service design, vol 1. Prentice Hall, Upper Saddle River

Farrar C, Worden K (2007) An introduction to structural health monitoring. Philos Trans Soc A: Math Phys Eng Sci 365(1851):303–315

FENICS (2014) http://www.fenicsproject.org. Accessed 22 Apr 2014

Furtner P, Stöger M, Schreyer M (2013) SHM DATA—management, treatment, analysis and interpretation—a solution for permanent monitoring systems. In: 6th International conference on structural health monitoring of intelligent infrastructure, Hong Kong, 9–11 Dec 2013

Gröger G, Kolbe TH, Nagel C, Häfele KH (2012) Open-GIS® city geography markup language (CityGML) encoding standard, version 2.0.0, OGC 08-007r2

INSPIRE directive (2007) Directive 2007/2/EC of the European parliament and of the council of 14 March 2007 establishing an Infrastructure for spatial information in the European community (INSPIRE). Online: http://www.eur-lex.europa.eu/LexUriServ/LexUriServ.do?uri=CELEX:32007L0002:EN:NOT. Accessed 22 Apr 2014

Jäger, R (1988) Analyse und Optimierung geodätischer Netze nach spektralen Kriterien und mechanischen Analogien, DGK, Reihe C, Nr. 342, München

Kunde F (2013) CityGML in PostGIS : Portierung, Anwendung und Performanz-Analyse am Beipiel der 3D City Database von Berlin, Masterarbeit am Institut für Geographie, Universität Potsdam, unveröffentlicht

Lienhart W (2007) Analysis of inhomogeneous structural monitoring data. Shaker Verlag, Aachen

Nagel C, Stadler A (2008) Die Oracle-Schnittstelle des Berliner 3D-Stadtmodells In: Clemen C (Hrsg) Entwicklerforum Geoinformationstechnik, Shaker Verlag, pp 197–221

Neitzel F, Weisbrich S, Wu CC (2014) Integration der finite-elemente-methode in die ausgleichungsrechnung zur parameteridentifikation. In: Wieser A (ed) Ingenieur-vermessung 2014, Beiträge zum 17. Internationalen Ingenieurvermessungskurs Zürich, Herbert Wichmann Verlag, pp 301–310

Ray S, Simion B, Brown AD (2011) Jackpine: a benchmark to evaluate spatial database performance. In: International conference on data engineering, IEEE

Resch B (2012) Standardisierte Geosensornetzwerke für Umweltbeobachtung in naher Echtzeit. http://www.berndresch.com/download/work/publications/resch_geosensornetzwer-ke_rt-gis_2013.pdf. Accessed 22 April 2014

Stadler A, Nagel C, König G, Kolbe TH (2008) Making interoperability persistent: a 3D geo database based on CityGML, In: Lee J, Zlatanova S (Hrsg) 3D Geo-information sciences, selected papers from the 3rd international workshop on 3D geo-information, Seoul, Korea. LNG&C Series. Springer, pp 175–192

Teskey WF (1988) Integrierte Analyse geodätischer und geotechnischer Daten sowie physikalischer Modelldaten zur Beschreibung des Deformationsverhaltens großer Erddämme unter statischer Belastung, DGK, Reihe C, Nr. 341, München

Welsch W, Heunecke O, Kuhlmann H (2000) Auswertung geodätischer Über-wachungsmessungen. In: Möser M, Müller G, Schlemmer H, Werner H (eds) Hand-buch Ingenieurgeodäsie. Herbert Wichmann Verlag, Heidelberg

Worden K, Dulieu-Barton JM (2004) An overview of intelligent fault detection in systems and structures. Struct Health Monit 3(1):85–98

Zienkiewicz OC (1971) The finite element method in engineering science. McGraw-Hill, London

# Requirements on Building Models Enabling the Guidance in a Navigation Scenario Using Cognitive Concepts

**Katrin Arendholz and Thomas Becker**

**Abstract** The increased growth of location-based services and the ongoing work on 3D GIS applications along with the fact that people spend most of their time indoors lead to a strong motivation for support and development in indoor space applications. When we ask someone the way, it is noticeable that people give route descriptions in an abstract and symbolic manner referring to landmarks or signs. Moreover, we know that people comprehend and structure space in recurring imaginative patterns—image schemata—while moving through and interacting with their environment. However, existing indoor navigation frameworks as well as building models do not integrate these cognitive concepts since mechanisms for defining such required semantics have not been developed yet. In order to support a navigation scenario in indoor environments by integrating human wayfinding principles in route instructions, we (1) identified potential landmarks based on cognitive concepts; (2) evaluated them according to a set of requirements; and (3) established a thematic framework for enhancing existing building models with the needed semantics and requirements. This framework will help developers to add the required attributes to the correct semantic classes of CityGML, IFC, or KML models to provide suitable routing instructions in terms of in-house routing out of these models.

**Keywords** Human wayfinding · Indoor route description · Image schemata · Landmarks · Building model

K. Arendholz (✉) · T. Becker
Institute for Geodesy und Geoinformation Science, Technische Universität Berlin,
Straße des 17. Juni, 10623 Berlin, Germany
e-mail: k.arendholz@gmail.com

T. Becker
e-mail: thomas.becker@tu-berlin.de

© Springer International Publishing Switzerland 2015 
M. Breunig et al. (eds.), *3D Geoinformation Science*,
Lecture Notes in Geoinformation and Cartography,
DOI 10.1007/978-3-319-12181-9_10

# 1 Introduction

Over the past decade car navigation systems became a standard since people trust in the provided information, whereas navigation systems within buildings have mostly been neglected (Brunner-Friedrich and Radoczky 2006; Lorenz et al. 2006). Main challenges in this research field comprise the use and development of proper positioning techniques, navigation models, and personalized guidance systems that are adapted to the user's preferences or needs. Initially, indoor navigation approaches were developed for rescue and evacuation applications (Lee and Zlatanova 2008). Nowadays, the focus is set on general navigation in complex indoor environments since people observed the convenience of location-based services (Yuan and Schneider 2010). Due to the facts, that people spend a lot of their time indoors and that an increasingly aging population requires more assistance, it is not difficult to predict that these approaches will become a prospective standard to guide people in unknown indoor environments such as hospitals, libraries, airports or shopping malls (Shayeganfar et al. 2008; Worboys 2011).

While conventional car navigation systems usually give turn-by-turn instructions, people tend to give route instructions in a symbolic manner often referring to landmarks, signs or symbols and they comprehend and structure their environment in an imaginative pattern, namely image schemata. "Image schemas are directly meaningful ("experiential"/"embodied"), preconceptual structures, which arise from, or are grounded in, human recurrent bodily movements through space, perceptual interactions, and ways of manipulating objects" (Hampe 2005). They are in current use for intuitive design issues in the field of human–computer interactions (Hurtienne 2011). However, it is also important to investigate people's perceptual and cognitive structures in order to be able to model their behaviour in spatial information systems, such as an indoor navigation application (Raubal and Egenhofer 1998).

The realization of an intuitive, landmark-based indoor navigation system requires appropriate data sources and building models. It is desirable to derive route description-relevant information (e.g. landmarks) directly from this base. However, existing 3D building models (e.g. CityGML) do not provide mechanisms to define navigation constraints or do not integrate the required entities and their semantics for supporting human wayfinding principles in any navigation scenario because the required entities are completely unknown. Furthermore, proposed indoor navigation frameworks, such as the Multilayered Space-Event Model (MLSEM) or IndoorGML respectively, do not aim to provide this kind of semantics, but rather prefer the integration of semantic information from suitable building models, e.g. CityGML or IFC (Becker et al. 2009; Nagel et al. 2010; Lee et al. 2014).

To include semantics that support the generation of human route instructions within an indoor navigation scenario, we investigated the influence of cognitive concepts (image schemata, landmarks, signs) on the human wayfinding process and we defined a minimum set of requirements that needs to be regarded.

This paper presents an approach how these requirements can be integrated in existing building models by establishing a thematic framework capturing these semantics. The interoperable-designed framework gives guidelines to developers how to enhance building models with the required semantics.

The structure of this paper looks as follows: Sect. 2 briefly outlines previous research activities and concepts in terms of pedestrian/indoor navigation and highlights a variety of existing indoor models. In Sect. 3 we identify suitable objects in indoor environments for a landmark-based guidance by investigating cognitive concepts (image schemata, landmark theory, and common-sense knowledge). Besides these concepts, more aspects—such as the landmark salience or user requirements—need to be considered to guarantee an appropriate landmark selection. Section 4 covers the identified requirements and presents our approach on how to enhance existing building models with these human wayfinding principles. Finally, in Sect. 5, we summarize this paper.

## 2 Related Work

Modelling an indoor navigation application differs from conventional car navigation systems. Different aspects—such as the positioning technique, user preferences, perceptual abilities of the user, or space modelling of a 3D environment—need to be reconsidered. The design of an intuitive and proper indoor navigation system is still an open research task.

Over the past years, research activities have mainly focused on the semantic modelling to support the wayfinding in built environments. Here, the literature often refers to ontologies, namely semantic models that have representational and reasoning capabilities. OntoNAV, ONALIN and C-NGINE are examples for these ontologies which propose guidelines how to model an indoor navigation system adapting to special needs, disabilities and preferences of the user (Anagnostopoulos et al. 2005; Dudas et al. 2009; Kritsotakis et al. 2009). A user-specific solution for blind and visually impaired people has been established by Swobodzinski and Raubal (2009).

In the context of indoor space representation, systematic approaches on constructing spatial models have been introduced. While Yuan and Schneider (2010) suggest a cube-based representation inspired by LEGO bricks, Lorenz et al. (2006) and Stoffel et al. (2007) present models consisting of structured graphs with nodes and edges carrying semantics to support a navigation scenario. To realize path-planning Lin et al. (2013) developed an extraction method based on IFC files as input. Another framework on space modelling and graph derivation has been developed by Nagel et al. (2010). Here, information can be derived from either IFC or CityGML building models.

The presented approaches on semantic modelling have in common that they adapt the path-finding algorithm to predefined rules in order to provide an optimal/best traversable route rather than the quickest/shortest route for the user. The frameworks on space modelling mainly focus on a correct representation of built environments concerning their geometric and topological characteristics in order to facilitate (personalized) path-planning.

However, little work has been done from the end user perspective concerning the presentation form of such a user-specific route. Although, several researchers have already revealed the benefit of integrating human wayfinding principles—such as the usage of landmarks—in route descriptions (Lovelace et al. 1999; Klippel and Winter 2005; Klippel et al. 2009; Radoczky 2007), the proposed indoor navigation models do not include such semantics and rely on existing base models (CAD drawings, building models etc.). Heiniz et al. (2012) propose a landmark-based concept to guide users through buildings by using images of the surroundings and textual instructions. However, their framework also focusses on the implementation of the different navigation modules (e.g. building graph, routing module etc.) and does not address the origin and data management of semantics that support the human wayfinding process.

# 3 Using Cognitive Concepts in Route Descriptions

This section gives an overview of the cognitive concepts and additional constraints that influence people's behaviour and landmark selection within a navigation task.

## *3.1 Identification of Objects Based on Cognitive Concepts*

To enhance route descriptions with human wayfinding principles means to understand how people acquire, structure, cognize and verbally present spatial knowledge within a navigation task. For this purpose, the concepts of image schemata, commonsense knowledge (e.g. signs) and landmarks were investigated. In this chapter, we show how these concepts can be used to identify landmarks in built environments.

### 3.1.1 Image Schemata

Image schemata are described as recurring, imaginative patterns of perceptual interactions that structure our experiences while operating beneath people's conscious awareness (Johnson 1987). Raubal et al. (1997) showed that their use is a powerful method to describe human spatial cognition related to navigation tasks. We assume that objects that share the characteristics of image schemata are more effectively cognized by people during the wayfinding process. Hence, such objects might serve as landmarks in indoor environments.

In order to detect relevant image schemata in the wayfinding process, we investigated different studies deducing image schemata from natural-language expressions (Raubal et al. 1997; Raubal and Egenhofer 1998; Adams 2000; Freundschuh and Sharma 1995; Frank and Raubal 1999).

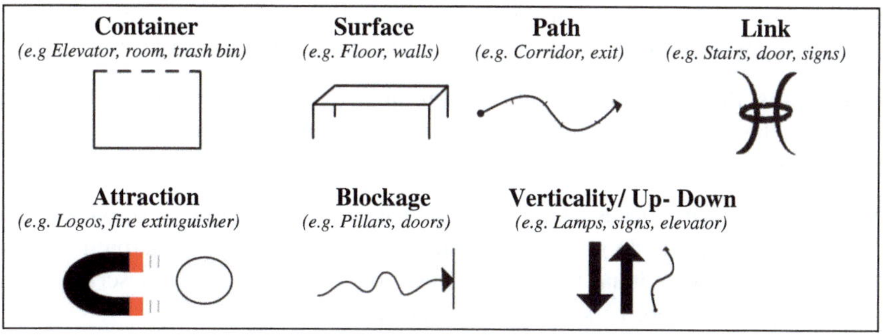


**Fig. 1** Images of the seven selected spatial image schemata with examples for corresponding objects [images were taken from IBIS-Projekt (2012)]

We evaluated seven spatial image schemata—**Container, Surface, Path, Link, Attraction, Blockage, Verticality/Up-Down**—from Raubal et al. (1997) as being highly relevant in the human wayfinding process and could detect corresponding objects in built environments (Fig. 1).

### 3.1.2 Common-Sense Knowledge (Signs)

As people give route instructions in a symbolic manner, they often refer to sign-posts, symbols, labels, logos, or pictograms becoming a landmark in their mental representation. They sense the sign; remember the meaning behind; and link it with the remembered action belonging to it. If they have never seen a certain sign before, people instantiate the newly gained knowledge. This procedure resembles the acquisition of image schemata and spatial knowledge respectively (Hurtienne 2011; Tomko 2007).

During the research, we investigated the following signs as being relevant as a landmark in indoor environments. Corresponding examples are given in Fig. 2.

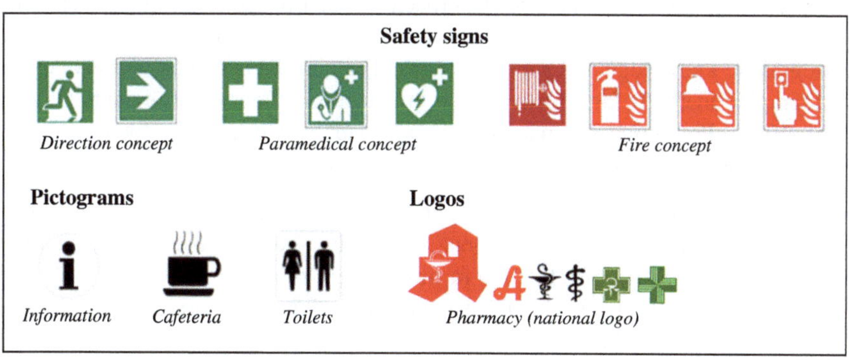


**Fig. 2** Safety signs according to DIN EN ISO 7010, pictograms indicating a room's function, and an example of a sector logo

**Safety signs** Safety signs can be found all over the building's interior and are mapped on evacuation plans. Their style is standardized in DIN EN ISO 7010.

**Pictograms** Pictograms often indicate a room's function, e.g. toilets or cafeteria. Although their style is not standardized, people understand their meaning applying common-sense reasoning.

**Name Plates** Name plates can also contribute to the wayfinding process with the restriction that people do not apply common-sense reasoning by interpreting signs but just reading them. It has to be drawn attention to the fact that the descriptions on the name plates cannot be sensed from a greater distance.

**Logos** Company or sector logos may also serve as landmarks in indoor environments as long as their locations stay permanently. However, it has to be distinguished between internationally and nationally accepted logos. While McDonald's is an internationally acting company that standardizes its products and advertising, logos applied in the pharmacy sector depend on the country (see Fig. 2). Hence, it is suggested to use internationally known logos to provide every wayfinder with the same degree of information.

### 3.1.3 Landmarks

A landmark is an important, significant and recognizable feature at an outstanding position which is used as reference point (Sorrows and Hirtle 1999; Raubal and Winter 2002; Klippel and Winter 2005; Brunner-Friedrich and Radoczky 2006; Klippel et al. 2009). Humans conceptualize landmarks either as point-like, linear or area-like entities (Klippel et al. 2009). Human route descriptions comprise the use of landmarks. They support the human wayfinding process by illustrating decision points or giving route verification in people's cognitive maps.

Annual user studies regarding different aspects of human wayfinding have been conducted during the "Long Night of Sciences" at the TU Berlin's main building (Lorenz et al. 2010; Lorenz and Thierbach 2012; Lorenz et al. 2013). The collaboration between the GIS department and the sociologists allows the bundling of cartographic knowledge and social research. The aim is to develop and investigate cartographic methods for effective route guidance in indoor environments (Lorenz et al. 2013). Presenting landmarks on paper maps was one of the focused design issues during the studies in 2011. The following landmarks were employed:

- Stairs, elevators, photo booth, memorial, cafeteria, trash bin, ladder, picture of TU Berlin, fire extinguisher, toilets, University shop, main entrance, audimax (a big lecture hall), info, H 2051 (a room label), propeller, university cash box/ pay office, view at the Geodätenstand, team civil engineering, astro-physical laboratory, tea kitchen.

The user studies are treated as social field experiments (factorial design without control group) comprised of an a-priori survey, a field experiment (navigate through

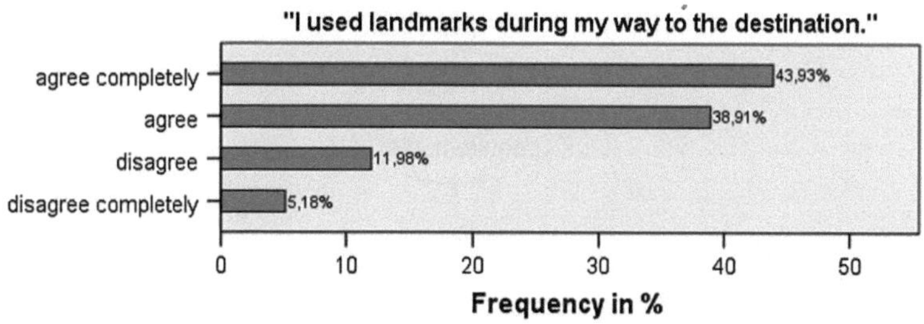


**Fig. 3** Participants' answers on their use of landmarks during the field experiment at the "Long Night of Sciences" 2011 in Berlin. This analysis of the a-posteriori survey is based on the survey type "adults" having 835 valid answers

the building) and an a-posteriori survey with open-ended questionnaire. In total, 2,400 visitors, aged between 4 and 78 years, attended the experiment, 1,140 of them only in the year 2011. Examining the a-priori and the a-posteriori surveys of 2011 helped to answer how people used and evaluated the employed landmarks on the given maps.

The results give positive feedback with respect to the utility of the employed landmarks. As can be seen in Fig. 3, over 80 % of the participants (adults) valued and used the provided landmarks. Moreover, further analyses prove that additional landmarks along the routes cause an improvement in the user's satisfaction.

In summary, analyses of the existing user study clearly show that landmarks represent a necessary element in route instructions. Based on these findings, we discovered the need to integrate landmarks and other description-relevant information in building models, so that this kind of information can be directly taken from the base of an indoor routing system.

## *3.2 Defining "Good" Landmarks for a Feasible Route Description in Indoor Environments*

The whole process of identifying suitable objects in indoor environments for a landmark-based guidance needs to be evaluated since every person perceives the environment individually. Applying "good" landmarks in navigation systems is therefore of great importance since poor chosen ones lead to user's confusion and increased workload (Burnett 2000). That is why more requirements need to be considered to get a reliable selection of landmarks. Therefore, our evaluation process regards the following aspects, depicted in Fig. 4:

1. The use of cognitive concepts to support human wayfinding principles,
2. salient characteristics so that an object can be identified as landmark by the wayfinder,
3. user-specific requirements that influence the choice of landmarks,
4. other, non-user-specific requirements that influence the determination of a "good" landmark.

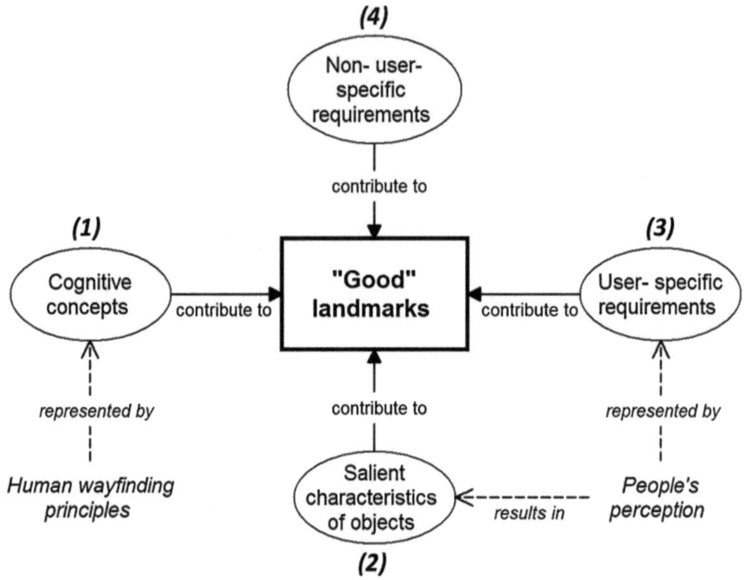


**Fig. 4** Criteria for the evaluation of a "good" landmark

### 3.2.1 Salient Characteristics of Landmarks

The quality of landmarks influences the user's satisfaction on finding the way to a destination. Within the literature this quality is named salience or "landmarkness" and guarantees a landmark's prominence and singularity (Burnett et al. 2001; Klippel and Winter 2005; Raubal and Winter 2002; Sorrows and Hirtle 1999). According to Sorrows' and Hirtle's (1999) classification, we distinguish between visual, cognitive and structural salience. A landmark will be stronger the more salient characteristics it possesses (Raubal and Winter 2002).

**Visual salience** A landmark's visual attractiveness can be achieved by certain visual characteristics that yield a sharp contrast with the surrounding. Raubal and Winter (2002) and Nuhn et al. (2012) already specified properties to determine the visual strength of a landmark—e.g. colour, shape, several façade properties. However, they refer to features in outdoor environments, especially buildings. We suggest the following visual characteristics as being relevant in an indoor navigation scenario: **Colour**, **material**, and **texture**. Investigations on the literature show that colour as a visual stimuli is strongly linked to the user's visual attention (Wolfe 2000). However, materials and textures—although visible by the sighted—effectively stimulate the tactile perception of blind and visually impaired people and are said to be important reference points (Passini and Proulx 1988). Hence, we propose to assign identified objects with these attributes in order to enhance the route descriptions, e.g. "turn right after passing the *red*, *brick-lined* wall".

**Cognitive salience** Significance due to content, meaning, use, cultural influence and prototypicality can convert an object into a cognitive landmark (Burnett et al. 2001; Sorrows and Hirtle 1999). We suggest including an object's **name**, **function** and **type** in existing data sources to enrich the quality of a landmark. However, we draw attention to the fact that cognitive landmarks tend to be personal (Sorrows and Hirtle 1999). Hence, they might be missed out by people being unfamiliar with the environment.

**Structural salience** Structural landmarks are attractive if they play an important role due to their prominent locations and high accessibility (Sorrows and Hirtle 1999). They help navigating especially in unfamiliar and known new environments. The structural salience can be determined by calculating the **node degree** of the landmark within an indoor network (Raubal and Winter 2002). The node degree is defined as the number of edges being adjacent to one node (Brunner-Friedrich and Radoczky 2006). Hence, a high node degree identifies a prominent location of a landmark.

### 3.2.2 User-Specific Requirements

Most current indoor navigation systems do not consider limitations of physically-, cognitively-, and sensory- impaired users (Dudas et al. 2009). We argue that the use of landmarks should be adapted to the users' needs. In our approach we consider, besides **non-handicapped people**, the following user groups:

- **Blind and visually impaired people**,
- **Wheelchair users**,
- **Children**.

Blind and visually impaired people have the highest demands since the salience and recognition of landmarks does not depend on visual characteristics anymore. They especially refer to textures stimulating their tactile perception. Wheelchair users and children are regarded due to other visibility conditions in comparison to averaged sized adults. Moreover, children's cognitive abilities differ from adult's ones (Heth et al. 1997).

### 3.2.3 Non-user Specific Requirements

Non-user-specific requirements comprise factors that may influence a wayfinder's ability to recognize or identify an object as landmark. Winter et al. (2005) revealed that the choice of landmarks also depends on the familiarity with the environment and illumination conditions. Considering indoor environments, we disregard illumination due to artificially created lighting conditions and we propose to use visual and structural landmarks since they are applied especially in unfamiliar environments. Furthermore, it seems likely that the bigger a landmark the more valuable it is since it can be seen from greater distances. We call this factor the **size–distance ratio** of an object. The ratio is defined by the size of an object in relation to its distance to the wayfinder and can be calculated as follows:

$$size - distance\ ratio = \frac{size\ of\ object}{distance\ between\ object\ and\ wayfinder}$$

Including the size–distance ratio would mean that landmarks will be adequately presented in a navigation system when people are able to recognize them from their location.

## 4 Using Cognitive Concepts in Building Models

This section presents the identified requirements which need to be included in existing building models to guarantee the generation of proper route descriptions. Furthermore, the established framework with its semantic classes and attributes is explained in detail.

### *4.1 Minimum Requirements*

In order to support a navigation scenario that follows human wayfinding principles, cognitive concepts, salient characteristics, user-specific and non-user-specific requirements need to be considered. Hence, integrating these requirements in proper building models helps to determine "good" landmarks for a proper indoor route description. The minimum requirements are presented below:

*Cognitive concepts*:

- A feature has an image schematic representation,
- A feature has a label/sign/logo,

*Salient Characteristics*:

- A feature is prominent in its location/has a significant node degree (structural salience),
- A feature's colour might be attractive to the wayfinder (visual salience),
- A feature's material/texture might be attractive to the wayfinder (visual salience),
- A feature's name, function or type might contribute to the wayfinding task (cognitive salience).

*User-Specific Requirements*:
A feature is sensed by different users:

- Adults without limitations,
- Wheelchair users,
- Children,
- Blind and visually impaired people.

*Non-User-Specific Requirements*:

- A feature's size (height and width) influences the visibility and recognition. The size-distance ratio needs to be calculated/evaluated.

## 4.2 Enhancing Building Models with Human Wayfinding Principles

In order to support the idea of being interoperable, we decided to establish a thematic framework that does not address explicitly other building models (Fig. 5). This framework shall provide semantic information of the indoor environment for enhancing route descriptions within indoor navigation applications. It shall give guidelines to developers on how to extend their models—such as IFC or CityGML —to enable human wayfinding from this base.

The framework's structure is designed in analogy to known class concepts (e.g. CityGML) and formally specified using principles of UML notation. However, it has to be emphasized that the displayed UML diagram is a thematic representation and cannot be used to generate xml schema files. In the following, a semantic definition of the features in Fig. 5 will be given:

***IndoorEnvironment*** characterizes the indoor environment of the whole building or parts of it.

***_AbstractLandmarkUnit*** represents the sum of all objects and spaces that resemble important landmarks in a built environment. These landmarks are further specialized to the following main subclasses: *_AbstractTransitionSpace*, *InstallationsAndFurniture*, and *Opening*.

***_AbstractTransitionSpace*** comprises features of greater extent that represent free space in a building and are generally used to traverse the navigable space within the built environment. This abstract class is further categorized into *VerticalSpace* and *HorizontalSpace*.

***VerticalSpace*** resembles indoor features referring to the vertical direction of locomotion in a 3D environment.

***HorizontalSpace*** describes features referring to the horizontal direction of locomotion in a 3D environment and represents the super class of *Room* and *Corridor*.

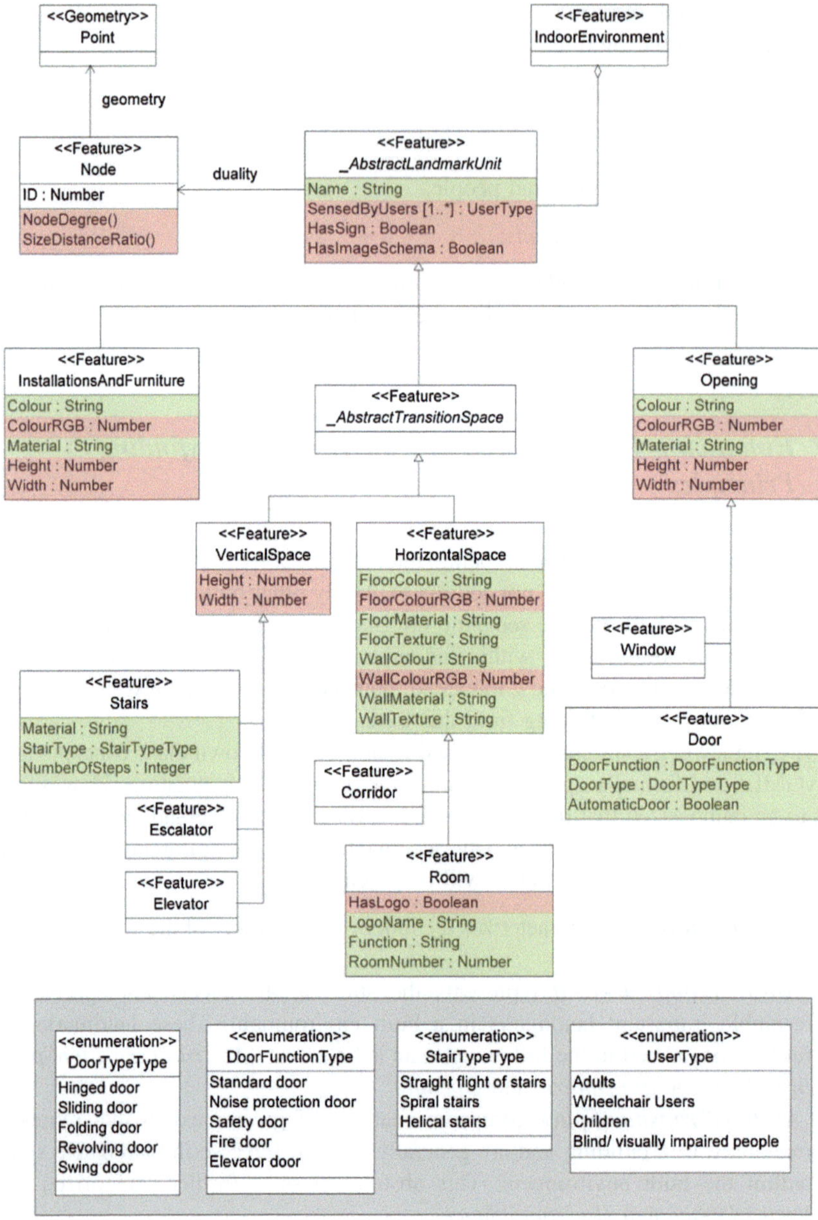


**Fig. 5** Thematic framework enabling guidance based on human wayfinding principles

***InstallationsAndFurniture*** defines all objects that can be found in a building and which neither belong to *TransitionSpace* nor *Opening*. The class' name is chosen as such since this model does not differentiate between movable (often indicated as furniture) or fixed objects (often indicated as installations) as proposed in other models related to indoor environments (Brown et al. 2013; Gröger et al. 2012; Hagedorn et al. 2009). Examples are fire extinguisher, trash bins, benches, or handrails.

***Opening*** is defined as an opening in a wall and is further categorized into *Door* and *Window*, quite similar to CityGML's Opening class.

***Node* and *Point***. The proposed thematic model does not set requirements on how to deal with geometry and topology of the features since it complies with existing building models' specifications and indoor navigation frameworks. However, to support the offered methods *NodeDegree()* and *SizeDistanceRatio()* and to simplify complex spatial relations, we suggest to map all landmarks in primal space—whether rooms, doors, trash bins or spaces—to Nodes (0D) in dual space by utilizing the Poincaré Duality concept (adapted from Nagel et al. (2010) with their proposal towards the MLSEM). Hence, the *Node* feature in dual space is associated with a *Point* as geometry.

In the following, a brief description of the attributes and operations shall be given. All the attributes' values are expressed in either textual, numeric or Boolean format as indicated in Fig. 5. The attributes' names are chosen in a conscious and self-explanatory manner. We distinguish between two types of attributes regarding their use within the framework.

The attributes—marked in green—are necessary to generate route instructions that resemble human route descriptions. Here, it is referred to the *Name* of the landmark, its texture (e.g. *FloorTexture*), *Material*, and *Colour*. In case of rooms in a building, people often mention the *RoomNumber* or the room's *Function*, the latter belongs to the category of a cognitive landmark. Moreover, a special function (e.g. *DoorFunction*) or type (e.g. *DoorType*) of a feature may contribute to a better route description.

The attributes—marked in red—are necessary to evaluate a feature according to the identified requirements so that it can become a potential landmark. The cognitive concepts are represented by the attributes *HasSign*, *HasImageSchema*, and *HasLogo*. The structural salience is induced by applying the *NodeDegree()* method and calculating the size distance ratio as representative for non-user-specific requirements can be performed by using the *Width* and *Height* attribute of features along with the *SizeDistanceRatio()* method. Selecting user-specific landmarks for the route instructions is realized by the attribute *SensedByUsers*.

***Name* and *LogoName*** are used to specify the landmark by its descriptive meaning or its labelling.

***SensedByUsers*** is used to identify the recognition of features dependant on different user groups and can occur multiple times. The discrete set of values, namely the user groups, is defined in the enumeration *UserType*.

***HasSign***, ***HasImageSchema*** **and** ***HasLogo*** specify if an object has a symbolic and/or image schematic representation or if a logo exists. If the Boolean attribute *HasLogo* is set to true, the route instructions should rather be generated by taking the landmark's *LogoName* instead of the *Name* since people are first attracted by big signs.

***Height*** **and** ***Width*** are needed to calculate the size of an object for performing the *SizeDistanceRatio()* method.

***Colour*** **and** ***Material*** are used to enhance the route description since people are more attracted by objects that differ from their surroundings.

***ColourRGB*** can be used to define the visual attraction (salience) of a landmark.

***Texture*** describes the surface structure of the entities and is aimed at a person's tactile and haptic perception in case of visual limitations (*FloorTexture*, *WallTexture*).

***Function*** expresses the intended purpose of the landmark, quite similar to CityGML's function type.

***Type*** is related to doors and stairs. While the *DoorType* defines a door's movement type, e.g. revolving door, the *StairType* is used to express the form of the stairs, e.g. straight flight of stairs.

***AutomaticDoor*** expresses a door's behaviour when approaching it. This information is very helpful for blind and visually impaired people since doors represent a risk for this user group.

***NumberOfSteps*** reflects the number of steps of a stair connecting two levels. This information is very helpful for blind and visually impaired people.

***RoomNumber*** is used to label a room by a number. This number can be found on name plates that are installed next to a room's entrance.

***NodeDegree()*** is intended to define the structural salience of a landmark. Therefore, a network comprised of nodes and edges needs to be generated.

***SizeDistanceRatio()*** needs the *Height* and *Width* attributes of the features for defining their sizes. The distance can be determined using the information from the edges of the geometric network that is generated for the routing task.

## 5 Conclusion and Outlook

Within an indoor navigation scenario, it is desired to consider people's spatial cognition as it is a fundamental process of acquiring spatial knowledge. The presented research contributes to the idea of enhancing route descriptions in indoor environments by human wayfinding principles. For this purpose, potential landmarks that follow these principles need to be identified, evaluated according to a set of requirements, and properly modelled so that they can be applied in an indoor navigation scenario. Albeit different models exist, that describe the interior of buildings very well, they do not contain semantics and features usable for guidance purposes. Thus, we presented an image schema- and salience-based framework for enriching those models to support guidance and the creation of route instructions out of the box. The framework introduces the semantic classes and attributes that people need to create a mental map and to be able to compare their reality against the virtual model. Moreover, it will help developers to add the required attributes to the correct semantic classes of CityGML, IFC, or KML models to provide suitable routing instructions in terms of in-house routing.

## References


Adams D (2000) Wayfinding tasks with image schemata and affordances during disasters. Executive analysis of fire department operations in emergency management. National Fire Academy, Kissimmee, pp 1–75

Anagnostopoulos C, Tsetsos V, Kikiras P, Hadjiefthymiades SP (2005) OntoNav: a semantic indoor navigation system. First workshop on semantics in mobile environments, Ayia, Cyprus, pp 1–8

Becker T, Nagel C, Kolbe TH (2009) A multilayered space-event model for navigation in indoor spaces. In: Lee J, Zlatanova S (eds) Advances in 3D geoinformation systems. Springer, Berlin, pp 61–77

Brown G, Nagel C, Zlatanova S, Kolbe TH (2013) Modelling 3D topographic space against indoor navigation requirements. In: Pouliot J, Daniel S, Hubert F, Zamyadi A (eds) Progress and new trends in 3D geoinformation sciences. Springer, Berlin, pp 1–22

Brunner-Friedrich B, Radoczky V (2006) Active landmarks in indoor environments. In: Bres S, Laurini R (eds) Visual information and information systems. Springer, Berlin, pp 203–215

Burnett G (2000) "Turn right at the traffic lights": the requirement for landmarks in vehicle navigation systems. J Navig 53:499–510

Burnett G, Smith D, May A (2001) Supporting the navigation task: characteristics of "good" landmarks. Contemp Ergonomics 1:441–446

Dudas PM, Ghafourian M, Karimi HA (2009) ONALIN: ontology and algorithm for indoor routing. In: MDM'09 tenth international conference on mobile data management: systems, services and middleware. IEEE, New York, pp 720–725

Frank AU, Raubal M (1999) Formal specification of image schemata—a step towards interoperability in geographic information systems. Spat Cogn Comput 1:67–101

Freundschuh S, Sharma M (1995) Spatial image schemata, locative terms, and geographic spaces in children's narrative: fostering spatial skills in children. Cartographica: Int J Geogr Inf Geovisualization 32:38–49

Gröger G, Kolbe TH, Nagel C, Häfele K-H (2012) OGC city geography markup language (CityGML) encoding standard, V. 2.0, Document No. 12-019

Hagedorn B, Trapp M, Glander T, Döllner J (2009) Towards an indoor level-of-detail model for route visualization. In: Tenth international conference on mobile data management: systems, services and middleware. IEEE, New York, pp 692–697

Hampe B (2005) Image schemas in cognitive linguistics: Introduction. In: Hampe B, Grady JE (eds) From perception to meaning: image schemas in cognitive linguistics. Mouton de Gruyter, Berlin, pp 1–14

Heiniz P, Krempels K-H, Terwelp C, Wuller S (2012) Landmark-based navigation in complex buildings. International conference on indoor positioning and indoor navigation (IPIN). IEEE, New York, pp 1–9

Heth CD, Cornell EH, Alberts DM (1997) Differential use of landmarks by 8-and 12-year-old children during route reversal navigation. J Environ Psychol 17:199–213

Hurtienne J (2011) Image schemas and design for intuitive use—exploring new guidance for user interface design. PhD thesis, TU Berlin, p 268

IBIS- Projekt (2012) Image_Schemata_Definitionen_Skizzen_Schlüsselwörter. Available at http://www.ibis-projekt.de/icc/assisto/med/8e2/8e210d57-442a-aa31-ec07-08366350fd4c,11111111-1111-1111-1111-111111111111.pdf. Accessed 6 April 2014

Johnson M (1987) The body in the mind: the bodily basis of reason and imagination. University of Chicago Press, Chicago, p 272

Klippel A, Winter S (2005) Structural salience of landmarks for route directions. In: Cohn G, Mark MD (eds) Spatial information theory—COSIT05 Elllicottville. Springer, Berlin, pp 347–362

Klippel A, Richter K, Hansen S (2009) Cognitively ergonomic route directions. In: Karimi HA (ed) Handbook of research on geoinformatics. Information Science Reference (IGI Global), Hershey, London, pp 230–238

Kritsotakis M, Michou M, Nikoloudakis E, Bikakis A, Patkos T, Antoniou G, Plexousakis D (2009) Design and implementation of a semantics-based contextual navigation guide for indoor environments. J Ambient Intell Smart Environ 1:261–285

Lee J, Zlatanova S (2008) A 3D data model and topological analyses for emergency response in urban areas. In: Zlatanova S, Li J (eds) Geospatial information technology for emergency response, vol 6. Taylor & Francis, London, pp 143–168

Lee J, Li K-J, Zlatanova S, Kolbe TH, Nagel C, Becker T (2014) OGC candidate IndoorGML Encoding Standard, V. 0.8.2, Document No. 14–005

Lin Y-H, Liu Y-S, Gao G, Han X-G, Lai C-Y, Gu M (2013) The IFC-based path planning for 3D indoor spaces. Adv Eng Inform 27:189–205

Lorenz A, Thierbach C (2012) Bewusst Wo? Gewusst Wie! Entwicklung innovativer kartographischer Methoden zur effektiven Navigation in Innenräumen. In: Weisbrich S, Kaden R (eds) Entwicklerforum Geodäsie und Geoinformationstechnik 2011. Shaker, Berlin, pp 89–100

Lorenz A, Thierbach C, Kolbe TH, Baur N (2010) Untersuchung der Effizienz und Akzeptanz von 2D- und 3D-Kartenvarianten für die Innenraumnavigation. In: Kohlhofer G, Franzen M (eds) Vorträge Dreiländertagung OVG, DGPF und SGPF - 30. Wissenschaftlich-Technische Jahrestagung der DGPF. Vienna, p 342–355

Lorenz A, Thierbach C, Baur N, Kolbe TH (2013) App-free zone: paper maps as alternative to electronic indoor navigation aids and their empirical evaluation with large user bases. In: Krisp JM (ed) Progress in location-based services. Springer, Berlin, pp 319–338

Lorenz B, Ohlbach HJ, Stoffel E-P (2006) A hybrid spatial model for representing indoor environments. In: Carswell JD, Tezuka T (eds) Web and wireless geographical information systems. Springer, Berlin, pp 102–112

Lovelace KL, Hegarty M, Montello DR (1999) Elements of good route directions in familiar and unfamiliar environments. In: Freksa C, Mark D (eds) Spatial information theory. Cognitive and computational foundations of geographic information science. Springer, Berlin, pp 65–82

Nagel C, Becker T, Kaden R, Li K-J, Lee J, Kolbe TH (2010) OpenGIS® requirements and space-event modeling for indoor navigation, Discussion Paper, OGC 10–191r1

Nuhn E, Munich U, Reinhardt W, Haske B (2012) Generation of landmarks from 3D city models and OSM data. In: Gensel J, Josselin D, Vandenbroucke D (eds) Proceedings of the AGILE'2012 international conference on geographic information science. Avignon, pp 365–369

Passini R, Proulx G (1988) Wayfinding without vision an experiment with congenitally totally blind people. Environ Behav 20:227–252

Radoczky V (2007) How to design a pedestrian navigation system for indoor and outdoor environments. In: Gartner G, Peterson MP (eds) Location based services and telecartography. Springer, Berlin, pp 301–316

Raubal M, Egenhofer M (1998) Comparing the complexity of wayfinding tasks in built environments. Environ Plann B 25:895–913

Raubal M, Winter S (2002) Enriching wayfinding instructions with local landmarks. In: Egenhofer MJ, Mark DM (eds) Geographic information Science. Springer, Berlin, pp 243–259

Raubal M, Egenhofer MJ, Pfoser D, Tryfona N (1997) Structuring space with image schemata: wayfinding in airports as a case study. In: Hirtle SC, Frank AU (eds) Spatial information theory a theoretical basis for GIS. Springer, Berlin, pp 85–102

Shayeganfar F, Anjomshoaa A, Tjoa A (2008) A smart indoor navigation solution based on building information model and google android. In: Miesenberger K, Zagler W, Karshmer A (eds) Computers helping people with special needs. Springer, Berlin, pp 1050–1056

Sorrows M, Hirtle S (1999) The nature of landmarks for real and electronic spaces. In: Freksa C, Mark D (eds) Spatial information theory, cognitive and computational foundations of geographic information science. Springer, Berlin, pp 37–50

Stoffel E-P, Lorenz B, Ohlbach HJ (2007) Towards a semantic spatial model for pedestrian indoor navigation. In: Hainaut J-L, Rundensteiner EA, Kirchberg M, Bertolotto M, Brochhausen M, Chen Y-PP, Cherfi SS-S, Doerr M, Han H, Hartmann S, Parsons J, Poels G, Rolland C, Trujillo J, Yu E, Zimányie E (eds) Advances in conceptual modeling–foundations and applications. Springer, Berlin, pp 328–337

Swobodzinski M, Raubal M (2009) An indoor routing algorithm for the blind: development and comparison to a routing algorithm for the sighted. Int J Geogr Inf Sci 23:1315–1343

Tomko M (2007) Destination descriptions in urban environments. PhD thesis, University of Melbourne, p 193

Winter S, Raubal M, Nothegger C (2005) Focalizing measures of salience for wayfinding. In: Meng L, Reichenbacher T, Zipf A (eds) Map-based mobile services, theories, methods and implementations. Springer, Berlin, pp 125–139

Wolfe J (2000) Visual attention. In: De Valois KK (ed) Seeing: handbook of perception and cognition. Academic press, San Diego, pp 335–386

Worboys M (2011) Modeling indoor space. In: Proceedings of the 3rd ACM SIGSPATIAL international workshop on indoor spatial awareness. ACM, pp 1–6

Yuan W, Schneider M (2010) Supporting 3D route planning in indoor space based on the LEGO representation. In: Proceedings of the 2nd ACM SIGSPATIAL international workshop on indoor spatial awareness. ACM, pp 16–23

# Context Aware Indoor Route Planning Using Semantic 3D Building Models with Cloud Computing

**Aftab Ahmed Khan, Zhihang Yao and Thomas H. Kolbe**

**Abstract** In recent years, an increasing number of route planning applications and services have been developed and were brought to the market for different specific use cases. Most of those products are based on the client-server framework that typically combines the database server with the application server for storing the relevant network model, performing the routing calculation, and sending the result back to the client. However, this client-server framework mostly restricts users from changing, modifying, or augmenting the network model with respect to their specific contextual routing requirements. This paper presents a new approach for context aware route planning through coupling of a multilevel cloud-based system architecture with the complex IndoorGML data model, which is an upcoming OGC standard for representing and exchanging semantics, geometry, and topology information of indoor 3D building models. The geometric and logical network model can be rapidly extracted from the IndoorGML data model within the framework of the Multilayered Space-Event Model (MLSEM). Unlike the classical two-tier client-server architecture, the proposed multilevel cloud-based system allows exporting and uploading the network model from a database server to the cloud services that serve as an intermediate system-level to make the exported network model modifiable over the Internet without altering the original data. All changes of the network model with respect to different specific events will be applied in real-time, and the corresponding route planning calculations can then be carried out at the client-side respectively.

**Keywords** Indoor route planning · CityGML · IndoorGML · Cloud computing · WebGIS

A.A. Khan (✉) · Z. Yao · T.H. Kolbe
Technische Universität München, Munich, Germany
e-mail: aftab.khan@tum.de

Z. Yao
e-mail: zhihang.yao@tum.de

T.H. Kolbe
e-mail: thomas.kolbe@tum.de

© Springer International Publishing Switzerland 2015

M. Breunig et al. (eds.), *3D Geoinformation Science*,
Lecture Notes in Geoinformation and Cartography,
DOI 10.1007/978-3-319-12181-9_11

# 1 Introduction

Indoor route planning has been widely investigated in Robotics, Computer Graphics, and Geographical Information Science for emergency evacuation, automation, and indoor navigation. Traditionally, most applications or services supplied for route planning are constructed by means of the client-server model that typically combines the database server with the application server for storing the relevant network model, performing the routing calculation, and sending the result back to the client. However, this client-server based approach mostly restricts users from changing, modifying, or augmenting the network model with respect to their specific contextual routing requirements. However, there is a strong demand of developing a systematic approach that allows users to customize the network model according to their particular context in order to obtain the desired route planning result without needing to alter the original data stored in the central database.

In the context of a Smart Campus project at the Technical University of Munich (TUM) a campus information system is currently being developed. This project is intended to create an integrated platform that provides benefits for managing all kinds of building information and supports for various application fields like indoor route planning. The IndoorGML data model, which allows to model and describe the geometric, topological and semantic information of the complex indoor environment, can be utilized as the information backbone stored in a central database for all indoor navigation aspects of the Smart Campus platform. Normally, the datasets stored in the database should remain unchanged to ensure stable database maintenance. However, for the individual route planning use cases the users always demand to customize the dataset to perform the routing calculations accordingly. For instance, in case of a conference on the campus, the local organizing committee may want to exclude some areas and hallways from route planning by modifying the original datasets through adding some obstacles into the network model. However, this is neither supported nor permitted by the facility management department, who maintains the central database. Furthermore, IndoorGML has a complex data structure, and it is therefore very difficult for the normal users to use and customize this data model.

In this work, we propose a specific three-tier cloud-based system architecture. Which on the one side facilitates the IndoorGML data model managed within a 3D geodatabase to represent the complex interior environment and supports to carry out complex reasoning tasks like the determination of routing plans according to different contextual requirements of different users. On the other side it provides an intuitive and simple user interface realized by a 3D webclient. Both levels are linked by a dedicated information and application layer which employs cloud computing to provide the possibility for the normal user to customize the network model of the building according to their specific needs without altering the original data. The proposed approach allows exporting and uploading simplified subset of the complex IndoorGML data model to the cloud services serving as an intermediate system-level to make the exported network model easily accessible and modifiable over the Internet and for performing context-dependent route planning. The results of the route planning calculations are visualized and can be explored in the 3D webclient in a highly intuitive and user-friendly way.

The rest of the paper is organized as follows: Section 2 discusses related work, Sect. 3 describes IndoorGML, Sect. 4 presents the system architecture to couple the IndoorGML database with a multilevel cloud-based system, Sect. 5 shows example scenarios and their implementation on TUM main building model, and Sect. 6 discusses conclusions.

## 2 Related Work

The notion of context plays a key role in the development of indoor navigation systems (Becker et al. 2009a, b). Contextual information defines as any information that relates and uses to enrich the knowledge about the user's state, his or her environment, and capabilities (Afyouni et al. 2014). Context changes with respect to the requirements of the specific application and the user's activities in the given environment. Context aware indoor routing which is comparatively a new area of research (for the last two decades) many projects and research works are carried out to facilitate the user for indoor route planning in various indoor applications, e.g., facility management, disaster management, etc. Some of the related work is discussed in the following.

The Cyberguide system is one of the first indoor guiding system to guide tourists through both indoor and outdoor environments (Abowd et al. 1997). The system is 2D map based and gives location information to user through displaying an arrow on a room map according to the user's location. Furthermore, the system is equipped with the information of interesting sights within the building, or pathways that the user can access and visit. The authors' main wish list to improve their system at that time was modifiable information base so the real contextual information can be collected from the user and give him response from the system in real time.

One of the early 3D indoor routing application is presented by Meijers et al. (2005). For the purpose of routing for evacuation the authors extracted the graph model of 3D building model and used Oracle Spatial 10g to store and manage geodata in the application. The application has limitations to consider real contextual situations, for example, cannot put a walking restriction on a specific room considering a specific situation. It only computes the shortest path from one place to the other or the exit areas. The application's limitation to modify graph model according to users' requirements make its usage limited to a few scenarios.

iNav an indoor navigation system is presented by Kargl et al. (2007) for the real time routing and navigation, which is based on client-server architecture. The navigation system can be used on different PDA devices and consists of many distributed web services. The system provides user with his location information and details of events occurring around him. The events are regularly updated by service providers. So, it makes the system more contextually aware. iNav has major performance issues and its restriction on the user to modify the base information according to his needs makes the system dependent on service providers, and flow of the contextual information (from service provider to user) become one sided in indoor space.

Using the similar approach of client-server architecture Inoue et al. (2008) have provided indoor mobile navigation system which has some main features those include providing the current position of the user on a 2D floor map, changing the floor map according to the user's position, and showing the routes from current location to destination. The system lacks flexibility in terms of dealing with contextual requirements of users (e.g. user cannot modify in floor map if he require to change).

Karimi and Ghafourian (2010) introduced an indoor application to consider different requirements of the users to facilitate indoor navigation. The application considers user's capabilities (checking mobility impaired or visually impaired) and provides graph of the building accordingly for the routing purpose. The application lacks flexibility, the graph model of the building which is the base information for all users cannot be modified by user according to his contextual needs.

Another 3D indoor routing application for the decision support in the emergency situations is presented by Schilling and Goetz (2010). The application utilizes a 3D building model in CityGML and uses client-server architecture. The application is implemented through three system domains; federal, regional, and local to help the clients or users in rescue operations. In case of fire eruption, it provides a local spatial map of danger zone on request of the client from the event location. The application has a major drawback that is users (rescue staff at the event location) cannot modify or create constraints on the building routes (graph models) which are affected by fire to restrict other users to navigate. A similar 3D indoor routing web application is developed by Goetz (2012) using crowdsourced (OpenStreetMap) indoor geodata. The system provides routing services for the static situations based on the pre-computed routes making system's application limited to some predefined scenarios. It also restrict users to make on-demand route or modify route considering user's contextual requirements.

Apart from above discussed research motivated indoor routing systems there are 2D visualization web-based indoor routing maps developed by commercial companies like Google Indoor Maps (GoogleIndoorMaps 2014).

From the above brief overview, it is apparent that there are several research motivated and commercial indoor navigation systems to facilitate user considering its contextual requirements. However, most of the above discussed indoor applications or navigation systems are based on 2D maps, although some of the systems are implemented on 3D building models. They operate on traditional client-server architecture which restrict end users to modify or change in 3D or 2D maps at their source according to his or her contextual requirements. Because the system always intends to ensure consistency in source geodata. Furthermore, most of the systems use only 2D visualization for the end user in contrast to the fact that 3D visualization is advantageous.

On the other hand, there is a new indoor representation model, i.e., IndoorGML which is based on Multi Layered Space-Event Model (MLSEM) (Becker et al. 2009a, b; IndoorGML 2014b). The MLSEM allows to represent different thematic decompositions of indoor space through multi space layers, e.g., sensor space, topographic space, etc. Further it provides technique to integrate different layers to utilize for indoor navigation or localization of the subject or object. As the multi-layers represent different themes of indoor space they collectively make a complex representation of a 3D building model represented in IndoorGML to understand and use for the normal user. Therefore, there is a need of an approach that should simplify, extract, and utilize the complex representation of 3D building model represented in IndoorGML for the normal user to use in different applications, e.g., context aware routing. In addition, it should enable users to modify or change the topographic space layer according to his contextual requirements without modifying the main sourced geodata, and visualize results in a 3D visualization tool.

## 3 IndoorGML Representing 3D Indoor Building Model

IndoorGML, a draft standard of the OGC represents and allows for exchanging of geoinformation that is required to develop and implement indoor navigation systems. It contains essential features that play a crucial role in indoor navigation. It is considered complementary to other 3D building modelling standards, e.g., CityGML and IFC in providing indoor spatial information concentrating on indoor navigation.

IndoorGML consists of two frameworks which are the Structured Space Model and the Multi-Layered Space-Event Model (IndoorGML 2014a). The Structured Space Model (SSM) explains how each space layer evolved systematically within four segments. The SSM subdivides 3D building models into four subdivided segments; primal space and dual space on the one hand, and geometry and topology on the other hand shown as in Fig. 1. The topological relationships between 3D (or 2D) spatial objects are represented in topology space. The 3D spatial objects (Cells) in primal space are transformed to nodes (OD) in dual space using the Poincare duality transformation. Similarly, the topological adjacency relationships between 3D objects which form the boundary geometry in primal space are transformed to edges (1D) in dual space. Moreover, the nodes and edges of the Node Relationship Graph (NRG) are called state and transition respectively. An adjacency graph is formed in dual space representing a specific contextual primal space, e.g., topographic or sensor space. Furthermore, based on semantic information the adjacency graph is transformed into connectivity graph. The connectivity graph forms a unique space layer that consists of nodes and edges geometries. As the indoor space can be thematically decomposed into different cellular spaces. For example, a corridor can be represented as topographic area while it is also represented as WiFi

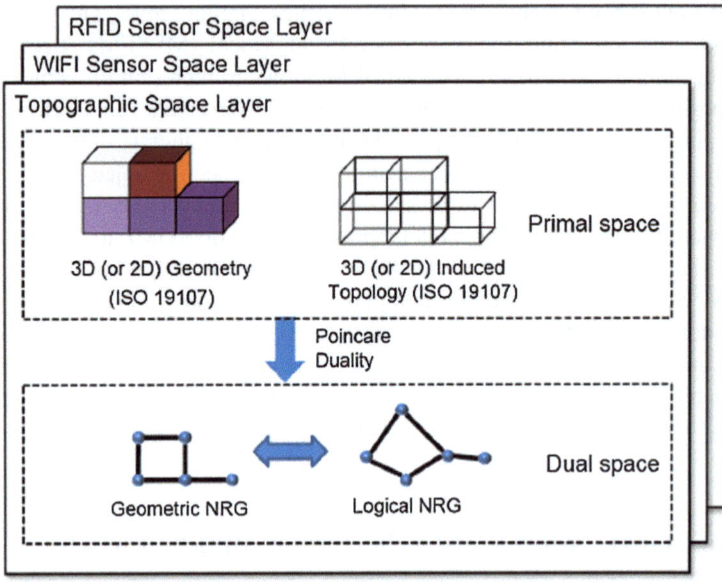


**Fig. 1** Multiple space layers and structured space model within each layer (IndoorGML 2014a)

coverage area and Bluetooth sensor coverage area. Each thematic interpretation area will form a different space layer in dual space through SSM architecture. This representation or the whole framework of multiple space layers is called Multiple Space-Event Layered Model (MLSEM). Further, the MLSEM provides method for integrating multiple space layers to support indoor location and information services shown in Fig. 2 (Becker et al. 2009a, b). IndoorGML also defines key concepts those include to reference any object in external datasets such as CityGML or IFC, connection with outdoor spaces, subspacing, and modularization to define extensions of IndoorGML to cover a specific thematic field.

Furthermore, the authors (IndoorGML 2014a, b; Nagel 2014) have provided details and application examples of IndoorGML. They further concluded that IndoorGML is a very flexible and have a sound mathematical foundation to represent and manage different thematic contexts of indoor semantic 3D building models for indoor localization and information services required for indoor navigation systems. But, the representation of a 3D indoor building model using IndoorGML containing different space layers to represent different thematic contexts of building with their mutual and hierarchical relationships makes a complex model for a normal user who may be only interested in one thematic context, e.g., indoor routing. Therefore, we extract the required information for indoor route planning and, thus, simplify the data structures for the end user to use for context aware indoor routing by coupling IndoorGML with a multilevel cloud-based system.

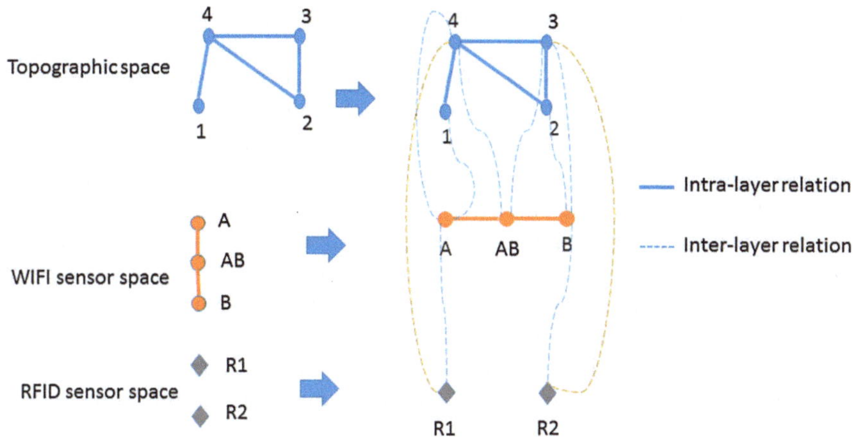


**Fig. 2** Multiple space layers in dual space (IndoorGML 2014a)

# 4 Coupling IndoorGML with Multilevel Cloud-Based System

As we discussed in Sect. 2 most of the indoor navigation systems or applications are constructed by the client-server architecture. The client-server based approach mostly restricts clients from modifying the base information on servers according to their specific contextual routing requirements. However, there is need of developing an approach that enable users to create subsets (e.g. routing network models) of the base information (e.g. main topographic network model) and modify in those subsets according to their specific contextual needs without changing in the original base information. Furthermore, the intended approach should use of new technologies based on the Internet that should enable users or clients to have a quick access to the stored data to address their real time contextual routing requirements. In addition, the normal user or the client should be able to access his relevant subset (network model) and give him a simple view from the complex base information (main routing network model) to deal with his contextual routing requirements. Considering these requirements, a new approach is developed through which a user will be able to create subgraphs of the main topographic network graph according to his contextual needs using a 3D user interactive and user friendly webclient. The subgraphs of the main network model can be modified, updated, and uploaded with the corresponding 3D building model to the cloud service to store in a Google Spreadsheet in real time. The uploaded subgraph and the corresponding 3D building model will be accessible to other users over the Internet instantly.

In recent years, the advent of cloud computing has enable to address the problems of resource scarceness, finite energy, and low connectivity (Satyanarayanan et al. 2009) on client devices to execute many useful programs that could aid the user and response to his queries in real time. Cloud computing is a combination of applications delivered as services and the hardware and systems software in the datacenters that provide those services over the Internet (Armbrust et al. 2010). Once these datacenters provides services to the general public in a pay-as-you-go manner, it becomes a public cloud. The main advantages of cloud computing are the availability of infinite computing resources on demand, enable cloud user to start at small scale, and increase resources only when there is an increase in their demand. In our proposed system we used a cloud service to facilitate the context aware indoor routing for the users.

The generic idea of coupling the IndoorGML data model with a multilevel cloud-based system to perform context aware indoor route planning is illustrated in Fig. 3. The whole system architecture consists of three tiers; information backbone, cloud service, and 3D webclient. In the information backbone tier, the semantic and geometric model of building are stored in an IndoorGML database, which allows to store and manage different contexts of indoor environment, as well as the mutual

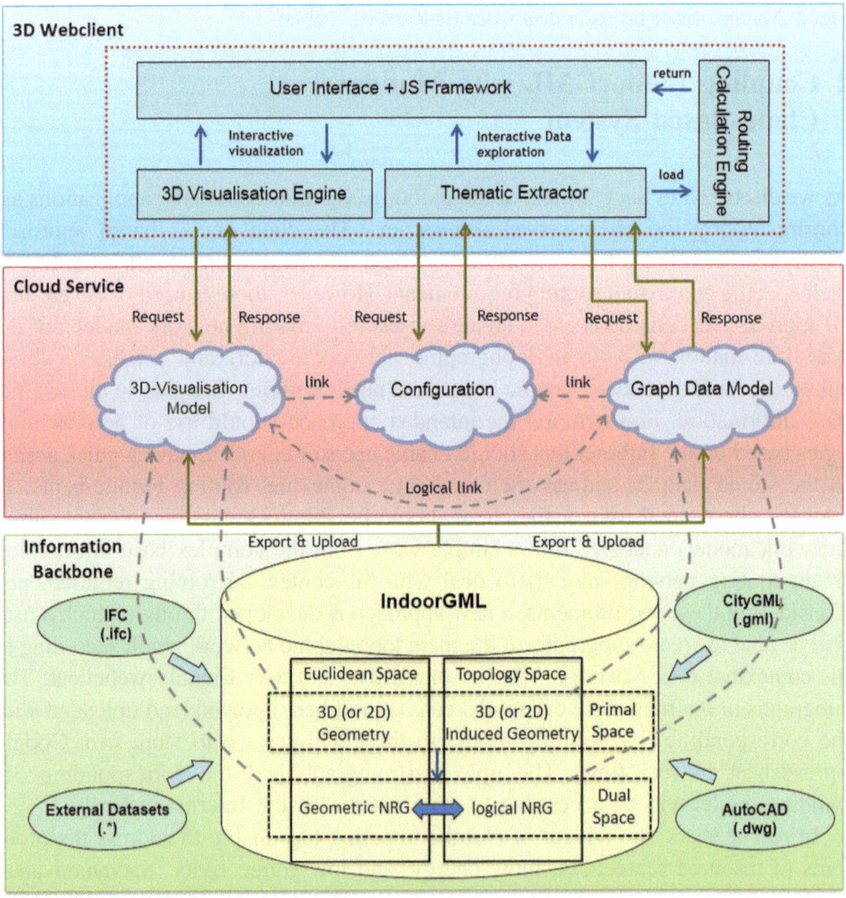


**Fig. 3** Specific three-tier system architecture for user-modifiable indoor route planning application

relationships of building parts. Other 3D data model standards like IFC, DWG, and CityGML, which are well-known in GIS, Architecture Engineering, and Construction (AEC) community, can be imported into the IndoorGML database. The network model can be extracted from the topographic space in IndoorGML using Multilayered Space-Event Model (MLSEM) method, which paves the way for the integration of multiple space layers such as topographic space, sensor spaces, and logical spaces to support navigation services (Becker et al. 2009a, b).

A specific 3D webclient acts as user interface to the end-user. It is developed to perform the functions such as route planning computation, interactive 3D visualization, and exploration. The main features and working architecture of the 3D webclient related to the context aware routing are explained in the following.

**Web-based and friendly user interface**: The client application is web-based and therefore available from any location with Internet access. Users can view the 3D webclient over the internet using a web browser to use directly without having to install any other software locally. It is a JavaScript-based static application and can operate with any webserver like Apache without the need of an application server which reduces the administrative effort. The basic structure of the user interface is created using the ExtJS JavaScript-based web framework. The 3D webclient enables to visualize graphical representation of the 3D building models and perform spatial operations such as geocoding. Furthermore, user can control the dynamic elements of 3D building models using JavaScript commands embedded with the Google Earth Plugin and the Google Maps API. Through an interactive 3D visualization a variety of features are available to display information of the target area. For example, panning, zoom, rotation with 3D view are provided by the Google Earth Plugin with its tools that provide the basic functions for navigation in the 3D map. In addition, the 3D webclient allows to select one or more objects and display their attribute values in a table. The selected objects can be both highlighted in the 3D view, as also be hidden from the current view.

**Interactive modification of thematic properties of the building model**: The editing feature of the webclient allows authorized users to change the thematic properties of the building model interactively for individual objects or entire groups of the selected objects (e.g. corrections, updating or adding more information). The edited property data is automatically stored in Google Spreadsheets.

**Context aware routing**: By means of this cloud-based system architecture it is possible to export arbitrary subgraphs of the main routing graph of the building model which are generated based on the different contextual requirements of the user and upload these to the cloud services to make them accessible over the Internet. Besides, a 3D visualisation building model linked with the exported subgraph can also be generated and exported in a similar way.

Furthermore, more than one pair of the exported datasets (graph data model and 3D visualisation model) can be grouped and referenced using one configuration document that allows users to control the distribution of different datasets and facilitates web applications to fetch sets of distributed datasets at once for speeding up loading time. The criteria to create subgraphs from the super graph of the building depends on the specific user and his authority to modify the main graph as well as his contextual requirements. The 3D webclient provides opportunity to each user to directly create constraints or edit attributes of building elements in an interactive way to create subgraphs which can be further uploaded to the cloud service to serve other users.

**Integration**: By using the Google Spreadsheet web application users can add more columns or other properties to the indoor objects. In addition, arbitrary KML files published through web or cloud services can be loaded into the webclient as a separate layer. This can either be a simple raster data such as OGC Web Map Services or 2D and 3D vector data.

**Queries**: Thematic inquiries are frequently used by analytical methods in GIS applications. In the webclient tool the building objects can be filtered by simple conditions on one or more attributes. In the described application context aware routing, for example, search for a specific type of rooms (e.g. lecture halls) or office of a specific person. Selected objects can be highlighted graphically and their relevant thematic properties can be displayed.

**Visualization of the topographic layer and the network model**: The main topographic space layer from IndoorGML building model is selected and uploaded as a layer into the cloud service to make it accessible over Internet. Furthermore, the corresponding network model from IndoorGML is uploaded as another layer to the cloud service. Both the 3D building model and the corresponding network model of the building with their semantic information are accessible from the webclient to visualize and analyse.

**Path computation and visualization**: The computation of the route plan from one room to the other amounts to the calculation of the shortest route between the two locations is performed directly in the webclient. In the application of the context aware routing, by simply selecting an initial room and target room the route for the user can be calculated. The result of the computed route will contain the list of rooms or corridors through which the person has to walk can be highlighted. As shown in Fig. 1, a 3D visualisation engine is embedded in the 3D webclient. It is responsible for the rendering of the visualisation model. Another client component named "thematic extractor" is involved to fetch and interpret the network model of the 3D building model stored in the cloud services. The users can utilize the "routing calculation engine" that performs the route calculation at the client-side with high performance due to the local caching of the network model. The results of route planning calculations can be directly visualized in the 3D webclient in an intuitive way.

## 5 Application Scenarios

The work flow supported by the mentioned system architecture in Sect. 4 is illustrated in Fig. 4. Users can be typically categorised into three groups: building administrator, scenario manager, and navigating user.

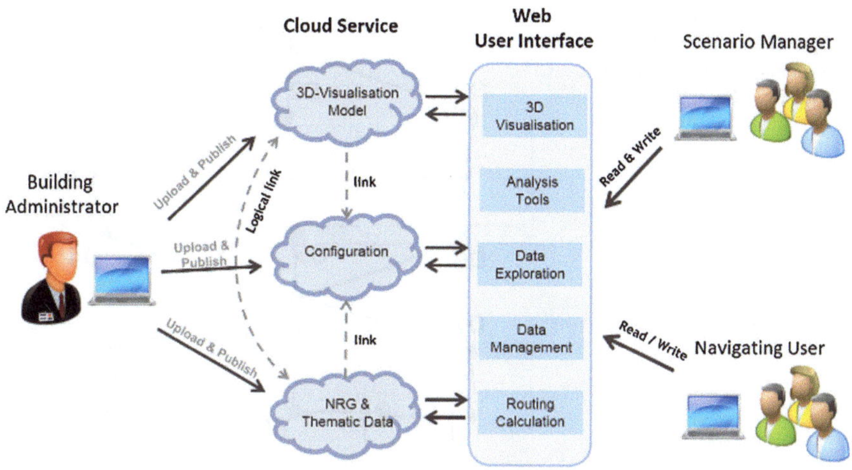


**Fig. 4** Application scenarios with three user groups with different access rights to perform functions on information backbone

The building administrator can export the configuration file associated with 3D visualisation models and network models, and upload to the cloud services to make them accessible from the 3D-webclient. These outsourced network models can be added, deleted, or modified by the authorized users (e.g. scenario manager) at any time without altering the original datasets stored in the central database (Herreruela et al. 2012). This strategy allows the scenario manager to modify and customize the network model in order to adapt it for the specific use cases. The modified network model can be fetched immediately from the cloud services by the 3D webclient for the route planning calculation. The desired route planning result will be intuitively presented to the navigating users via the 3D webclient. Since the cloud services provide support for access control, privileged navigating users or user groups can also be authorized by the scenario manager or the building manager, so that they are able to modify the network model by means of the functionalities shipped with 3D webclient such as exploration, query, and aggregation. On the other hand, since more than one outsourced dataset can be grouped and referenced in a separate configuration document, it is therefore possible to handle several application scenarios within one web application session, and one web application instance can also be used by more than one user or user group in turn. Example scenarios and screen shots of the results are given in the following paragraphs.

**Example scenario 1**: Consider a scenario in the main building of Technical University of Munich, where a GIS conference is going to be organized between dates 10–12 July, 2014. To facilitate conference participants with accurate indoor routing according to their specific requirements the building administrator assigns the task to the scenario manager and provides him the main topographic model of the building with the network model shown in Fig. 5 (2D view of a building). The scenario manager studies the conference plan and sessions' schedule and came to

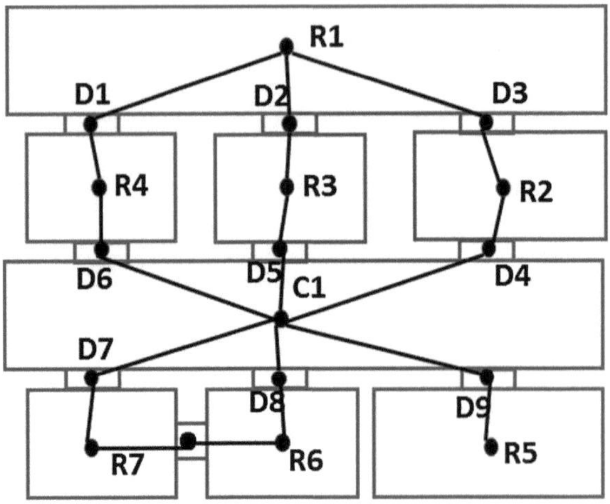


**Fig. 5** 2D view of the topographic model and the network model of the building provided by the building administrator

the decision that for two days room R3 and R4 must be closed to walk through by all participants because those are booked for private discussions. Therefore, he makes those rooms inaccessible (blocked) for all participants by editing the graph directly using the webclient and upload the new network model shown in Fig. 6 to the cloud service to access for participants so whenever they compute the route plan then those two rooms will be inaccessible and will not appear in their route computation by the system. Now the participants will be able to compute routes with 3D visualization of the building model considering the specific context (in this case without disturbing the private discussions in room A and B) during the conference.

**Example scenario 2**: In continuation of scenario 1, the conference has many sessions and each session is chaired by a session chairperson. Scenario manager provides the network model shown in Fig. 6 to the each session chairperson with

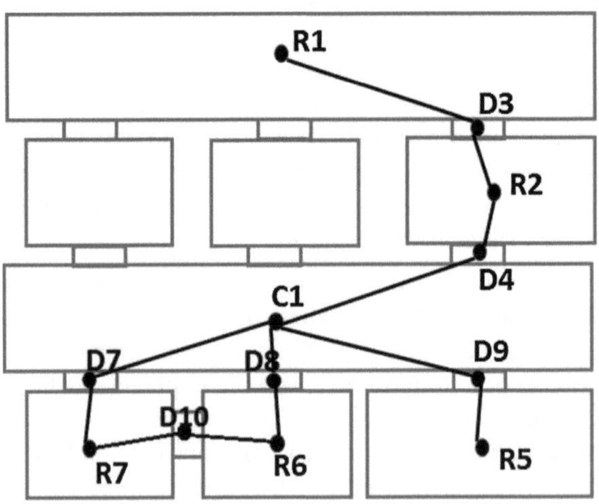


**Fig. 6** 2D view of the topographic model and the network model of the building after blocking rooms R3 and R4 by the scenario manager

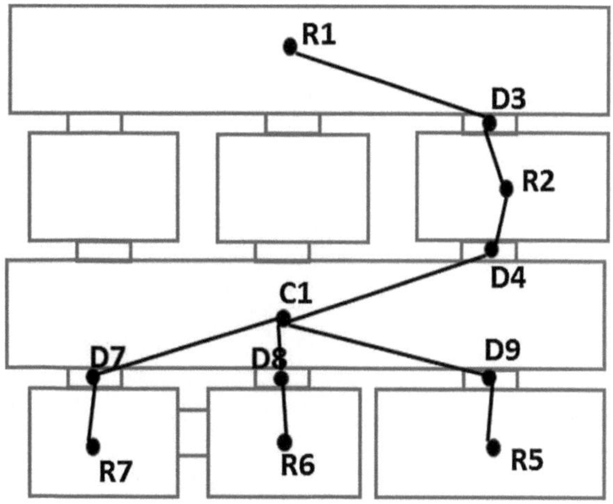


**Fig. 7** 2D view of the topographic model and the network model of the building after blocking door D10 by the session chair

permission to modify it according to specific requirements for the session participants. The conference has a visualization theme and its session is chaired by Mr. Yao in room R7 which has two entry doors D7 and D10. After having details of the conference sessions, and considering the requirements of the session participants he came with idea to close or block the door D10 for participants because it would be disturbing for the other session in room R6 that will be in progress during the closing and starting of his session. So participants should not walk through R6 to reach R7. Considering this visualization session's contextual requirement Mr. Yao modifies the network model of building and he blocks D10 for the participants of his session. He uploads the modified network model and the 3D visualization model of the building as shown in Fig. 7 to the cloud service so that the participants will get adapted routes without disturbing the session in R6.

The system architecture and example scenarios discussed in Sects. 4 and 5 are realized in the context of a Smart Campus project at Technical University of Munich (TUM) where a campus information system is currently being developed. This project plans to create an integrated platform that provides benefits for managing all kinds of building information. It will facilitate personnel of different departments and will support for various application fields like indoor route planning. The visualization of the 3D model of the main building of TUM and its network model in the webclient are shown in Figs. 8 and 9 respectively. From Fig. 8 it can be observed that apart from the 3D visualization of the building model the webclient provides a user friendly user interface to visualize all the attributes of each element of building which can be also edited by user. In Fig. 9 the network model or dual space of building is given where each node and edge represent room space cell and boundary cell of the building model respectively from primal space. Figure 9 further shows, webclient gives direct access to the user to interact and modify network model based on attributes of building model. In Fig. 10 the route plan (list of room numbers to go through and rooms are highlighted in yellow color)

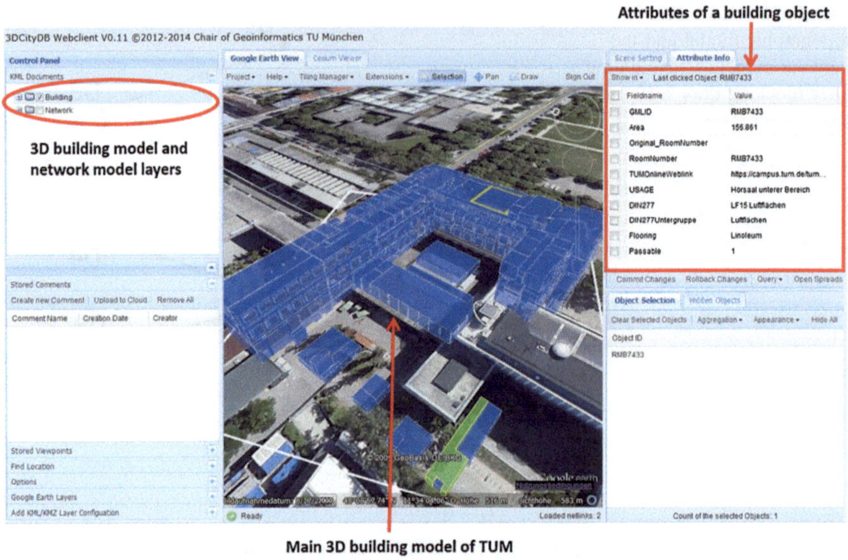


**Fig. 8** Excerpt of the 3D model of the main building of TUM in the 3D webclient (only interior rooms are shown)

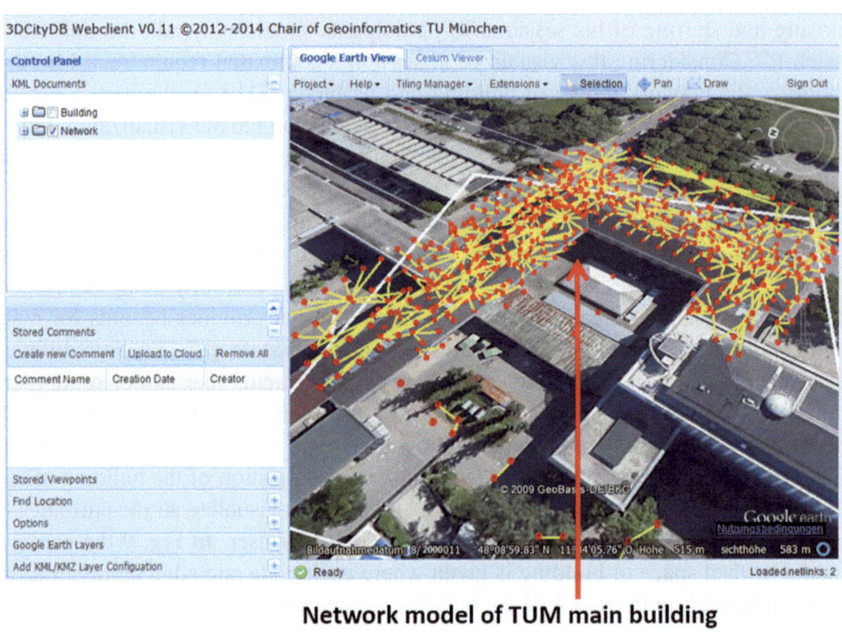


**Fig. 9** Network model layer of TUM main building in webclient

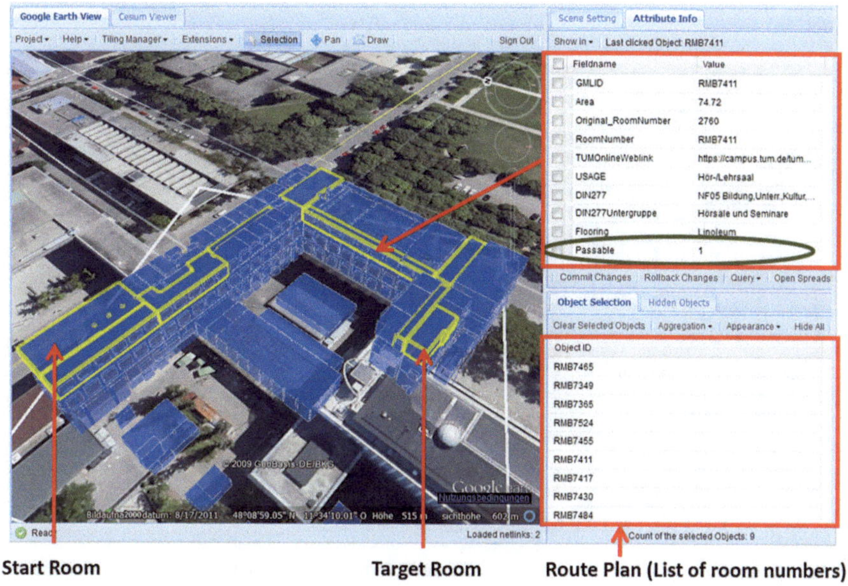


**Fig. 10** Route plan with all the semantic and geometric information of the route from the start room to the target room based on the main network model provided by the building administrator

for the user is computed through Dijkstra's algorithm by selecting two rooms (start and target room) of building based on the main topographic layer (network model provided by the building administrator explained in example scenario 1 shown in Fig. 9). Figure 11 shows the computed route plan (rooms are highlighted in yellow color) for the user by selecting the same two rooms (start and target selected earlier in Fig. 10) after putting restriction on room no RMB7411 to pass through for users by scenario manager due to some construction work. It can be observed that the rout plan in Fig. 10 differs from the route plan shown in Fig. 11 due to the fact that the users were provided two different network models by building administrator and scenario manager.

Furthermore, the webclient gives opportunity to users (building administrator, scenario manager, session chairperson, or normal user) according to their accessibility and modification rights to put constraints on the building elements through different attributes, for example, based on flooring type, usage, area, etc. and generate corresponding network model to upload into cloud service to use for a user or user group. The constraints applied through this webclient on the indoor space can also be used for the subspacing according to different locomotion types (walking, driving, flying) using their specific constraints (Khan and Kolbe 2012, 2013).

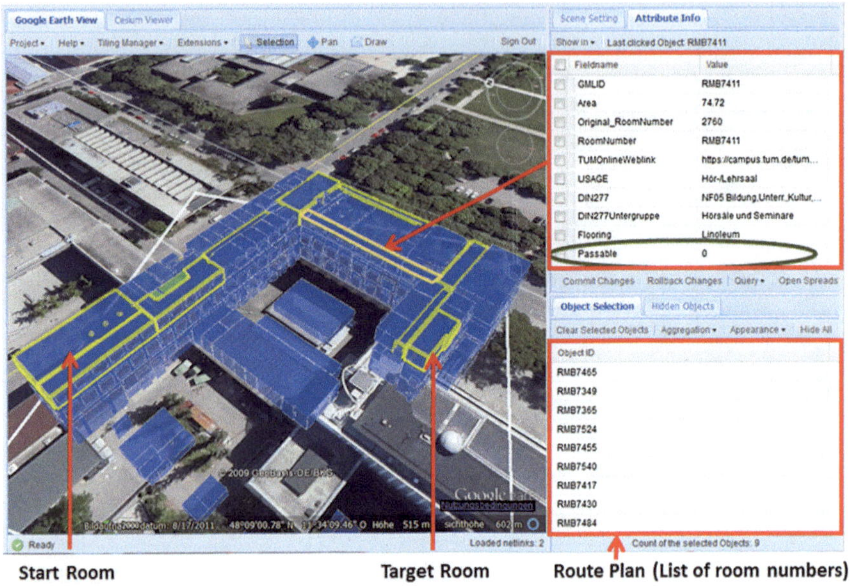


**Fig. 11** Route plan for the user from start to target room based on the network model provided by the scenario manager after restricting room no RMB7411 to pass through

## 6 Conclusions

In this paper we have presented a structured approach to combine two frameworks (multilevel cloud-based system and IndoorGML) to facilitate users for context aware indoor route planning. Based on IndoorGML, the network model along with the semantic and geometric information of the building can be completely stored in an IndoorGML database, which allows to carry out complex analyses on the one side and to generate and export arbitrary subsets of the original network model on the other side. The proposed system architecture utilizes cloud services to store the exported network model that can be dynamically customized and applied in different scenarios and disciplines for the route planning calculation, whose outcomes can be visualized and interactively explored via a specific 3D webclient. There are three groups of users (building administrator, scenario manager, and navigating user), and access rights are set accordingly to the role. Moreover, the presented approach coupling the IndoorGML data model with the specific cloud-based system, can also be investigated and applied for other applications that deal with other thematic contexts of the indoor environment based on network model, e.g. sensors' covering areas, etc. and their analysis.

The proposed system architecture also can be used to model navigation constraints generating from indoor space according to different contextual requirements, which can be directly stored in IndoorGML database by taking the input from user through webclient interface. The constraints stored in IndoorGML database given by the user through webclient will support the different types of locomotion in their navigation of the indoor space. It can support the navigation in outdoor environments provided if the system is connected with outdoor navigation model (e.g. to determine if a building is possible to navigate for the specific locomotion type from outdoor).

## References


Abowd GD, Atkeson CG, Hong J, Long S, Kooper R, Pinkerton M (1997) Cyberguide: a mobile context-aware tour guide. Wirel Netw 3(5):421–433

Afyouni I, Ray C, Claramunt C (2014) Spatial models for context-aware indoor navigation systems: a survey. J Spatial Inf Sci 4:85–123

Armbrust M, Fox A et al (2010) A view of cloud computing. Commun ACM 53(4):50–58

Becker T, Nagel C, Kolbe TH (2009a) A multilayered space-event model for navigation in indoor spaces. In: Lee J, Zlatanova S (eds) 3D geo-information sciences. Lecture notes in geoinformation and cartography. Springer, Berlin, pp 61–77

Becker T, Nagel C, Kolbe T H (2009b) Supporting contexts for indoor navigation using a multilayered space model. In: 10th International conference on mobile data management: systems, services and middleware, Taipei, May 2009. IEEE, pp 680–685

Goetz M (2012) Using crowdsourced indoor geodata for the creation of a three-dimensional indoor routing web application. Future Internet 4(2):575–591

GoogleIndoorMaps (2014) Google Indoor Maps. https://www.google.com/maps/about/partners/indoormaps/. Accessed 13 June 2014

Herreruela J, Nagel C, Kolbe TH (2012) Value-added services for 3D city models using cloud computing. In: Löwner M, Wohlfahrt R, Hillen F (eds) Mobilität und umwelt, proceedings of geoInformatik, Braunschweig, March 2012. Shaker Verlag, Aachen

Hijazi I, Ehlers M (2009) Web 3D routing in and between buildings. In: Proceedings of the 4th national GIS symposium, Al-Khobar, Saudi Arabia, 4–6 May 2009

Inoue Y, Ikeda T, Yamamoto K, Yamashita T, Sashima A, Kurumatani K (2008) Usability study of indoor mobile navigation system in commercial facilities. In: Proceedings of the 2nd international workshop on ubiquitous systems evaluation (USE'08), Seoul, Korea, 21 Sept 2008

IndoorGML (2014a) Open geospatial consortium (OGC) indoorGML draft. OpenGIS specification. OGC's document no. OGC 14-005r1, Version. v.0.9.0.

IndoorGML (2014b) www.indoorgml.net. Accessed 13 June 2014

Kargl F, Geßler S, Flerlage F (2007) The iNAV indoor navigation system. In: Ichikawa H, Cho W, Satoh I, Youn H (eds) Ubiquitous computing system. Lecture notes in computer science, vol 4836. Springer, Heidelberg, p 110–117

Karimi HA, Ghafourian M (2010) Indoor routing for individuals with special needs and preferences. Trans GIS 14(3):299–329

Khan AA, Kolbe TH (2012) Constraints and their role in subspacing for the locomotion types in indoor navigation. In: Proceedings of the international conference on indoor positioning and indoor navigation (IPIN), Sydney, Nov 2012. IEEE, pp 1–12

Khan AA, Kolbe TH (2013) Subspacing based on connected opening spaces and for different locomotion types using geometric and graph based representation in multilayered space-event model (MLSEM). In: ISPRS annals of photogrammetry, remote sensing and spatial information sciences, 13(II-2/W1), pp 173–185. doi: 10.5194/isprsannals-II-2-W1-173-2013

Lee J, Zlatanova S (2008) A 3D data model and topological analyses for emergency response in urban areas. In: Zlatanova S, Li J (eds) Geospatial information technology for emergency response. Taylor & Francis Group, London, pp 143–165

Meijers M, Zlatanova S, Pfeifer N (2005) 3D Geo-information indoors: structuring for evacuation. In: Proceedings of the 1st international ISPRS/EuroSDR/DGPF-workshop on next generation 3D city models (EuroSDRBonn), Bonn, 21–22 June, p. 6

Nagel C (2014) Spatio-semantic modelling of indoor environments for indoor navigation. Ph.D. thesis. Institute of Geodesy and Geoinformation Science, Technical University Berlin, Germany (in press)

Satyanarayanan M, Bahl P, Caceres R, Davies N (2009) The case for VM-based cloudlets in mobile computing. Pervasive Comput IEEE 8(4):14–23

Schilling A, Goetz M (2010) Decision support systems using 3D OGC services and indoor routing —example scenario from the OWS-6 testbed. Paper presented at the 5th 3D geoinfo conference, Berlin, Germany

Yao Z, Sindram M, Kaden R, Kolbe TH (2014) Cloud-basierter 3D-Webclient zur kollaborativen Planung energetischer Maßnahmen am Beispiel von Berlin und London. In: Kolbe TH, Bill R, Donaubauer A (eds) Geoinformations system, proceedings of the Münchner GI-Runde, 24–25 Feb 2014, Munich, pp 40–52

# Exploring the Benefits of 3D City Models in the Field of Urban Particles Distribution Modelling—A Comparison of Model Results

**Yahya Ghassoun, Marc-O. Löwner and Stephan Weber**

**Abstract** We present a comparison of a particles distribution model using 3D parameters derived from a CityGML-based 3D city model with an already advanced but 2D-based Land Use Regression model. Particles, especially ultrafine particles have significant influence on the health status of the urban population. Next to emission by cars and others, its distribution is tightly coupled to the local wind field and, therefore, to urban morphology influencing this wind field. However, 3D city models, especially CityGML have been almost ignored when modelling urban particles distribution. We introduce 3D parameters derived from a CityGML-based 3D city model in an already tested Land Use Regression model and explore the benefits of 3D city models in the field of particles distribution modelling, especially, by minimizing the number of parameters entered to the model and the good results that it has shown and explore the enhancement by combining both models.

**Keywords** Land use regression · CityGML · Ultrafine particle · 3D city model · Geostatistical model

## 1 Introduction and Problem Statement

Air quality in urban areas has a major influence on health of urban population. Health effects of ultrafine particles have been widely discussed and demonstrated in epidemiological and toxicological studies (WHO Panel 2013). The pollution effects

Y. Ghassoun (✉) · M.-O. Löwner
Institute for Geodesy and Photogrammetry, Technische Universität Branschweig, Brunswick, Germany
e-mail: y.ghassoun@tu-bs.de

M.-O. Löwner
e-mail: m-o.loewner@tu-bs.de

S. Weber
Institute of Geoecology, Technische Universität Branschweig, Brunswick, Germany
e-mail: s.weber@tu-braunschweig.de

© Springer International Publishing Switzerland 2015

M. Breunig et al. (eds.), *3D Geoinformation Science*,
Lecture Notes in Geoinformation and Cartography,
DOI 10.1007/978-3-319-12181-9_12

of ultrafine particles (UFP) on the health in urban area have increased mortality and morbidity especially from cardiovascular and respiratory diseases (Brook et al. 2010). Ultrafine particles with diameter less than 0.1 microns often are the direct product of the combustion of fossil fuels by road transport (Geiser et al. 2005). In addition, Hoffmann et al. (2006) showed that long term exposure to fine particulate matter (PM2.5) may accelerate the development and progression of atherosclerosis. The UFP concentration can be estimated from measurements of the particle number size distribution. At street canyon or near-traffic sites, the number of UFP generally accounts for the majority of total particle number concentrations, i.e., greater than 80–90 % (Morawska et al. 2008; Weber et al. 2013).

Since high traffic intensity is considered as an essential source for ultrafine particles in many cities, the proximity of the residence to roads was identified as a major determinant of health effects by particulate air pollution (Jerrett 2011). Thus, accurate assessment of exposure to traffic-related air pollution is critical for both, epidemiologic studies as well as for the estimation of health impacts of current and future road networks (Gilbert et al. 2005).

Urban complexity, i.e., the road network with different traffic intensity on road sections in combination with the three-dimensional structure of buildings, which influence climate and dispersion, leads to a specific spatial distribution pattern of urban UFP concentrations. Therefore, detailed assessment of exposure by measurement and modelling of fine dust distribution is necessary in the field of urban planning, traffic management and city system modelling.

Urban morphology can be analysed concerning geometrical properties of street canyons (Vardoulakis et al. 2002), building density, alignment of streets towards the prevailing wind direction, and characteristics of crossing sections. However, the latter is only little noticed when considering the redistribution of air and exchange of fine particles between different street systems (Brand and Löwner 2014).

Land Use Regression (LUR) models were developed as an alternative to dispersion models. As a geostatistical method, they can be applied for predicting local variation in traffic pollution and to obtain accurate, small scale air pollutant concentrations without a detailed pollutant emission inventory (Briggs et al. 2000; Brauer et al. 2003). LUR predicts pollution concentrations at a given site based on surrounding spatial characteristics and other identified auxiliary data which may act as a pollution predictor (Jerrett 2011).

LUR models are multiple linear regressions that assume independent residuals and use GIS-based covariates to predict concentrations at locations without measurements (Mercer et al. 2011). They have been widely applied in cities of North America and Europe (Saraswat et al. 2013). Henderson et al. (2007) developed LUR models for UFP in the city of Amsterdam including traffic intensity of nearest road, inverse of distance squared, population density and land use. Arian et al. (2007) included wind flow effects in LUR models for predicting $NO_2$ concentrations for health exposure studies in the greater Toronto-Hamilton region.

In comparison with meteorological data the main advantage of LUR is the incorporation of extensive data that can be measured with small errors (Jerrett et al. 2006). However, it neglects the process-oriented view of fine dust production, dilution or dispersion and accumulation or deposition. Only statistical measures are used to develop the prediction model without asking for causal linkage.

Recently, only few studies included 3D data of buildings and street canyons into LUR modelling. Tang et al. (2013) has used building height and geometry to enhance the representation of land use and the dispersion field in LUR. Eeftens et al. (2013) described an approach to calculate four indicators (the maximum aspect ratio, the mean buildings angle, median building angle, which is the angle between the horizontal street level and the line of sight to the top of surrounding buildings, and SkyView Factor) of urban canyon effect using 3D building data and evaluated the improvement on $NO_2$ and $NO_x$ LUR models in the Netherlands. Brauer et al. (2003) added 3D parameters (sampling height, street type, and canyon) as additional variables into the regression models to determine the extent to which prediction could be improved.

Today, 3D city models are increasingly available due to the standardization processes like CityGML (Gröger et al. 2012) for semantically enriched virtual 3D city models. An advantage of CityGML is its scalability to the requirements of the user and the data available (Löwner et al. 2013b). Today, they already support a lots of application fields like urban planning and hazard assessment (Löwner et al. 2013a). If only geometrically models are needed, KML models may be sufficient.

Here, we present a comparison of a UFP model using 3D parameters derived from a CityGML-based 3D city model with an already advanced but 2D-based Land Use Regression model. A land use regression based geostatistical model (LUR) is developed to estimate UFP concentrations at given locations using 2D and 3D parameters. Conceptually, both models were built to depict particle concentrations in urban microenvironments as the result of a 'process chain' that can be subdivided into three parts: emission, dilution/dispersion and deposition. All these processes take place under certain boundary conditions which are variable due to anthropogenic behaviour, weather conditions and the surrounding built environment. Therefore, the models are implemented with respective input parameters that are related to the three processes of production, dilution and accumulation. Incorporating 3D attributes to further develop the aforementioned LUR model for the prediction of UFP distribution within a city quarter leads to a more precise UFP estimation.

## *1.1 Study Area and Measurements*

The local scale variation of UFP is studied in the city of Braunschweig, a medium-sized city which is located in the south-eastern part of the federal state of Lower-Saxony, Germany. The 1.0 $km^2$ study area is characterized by street canyons with low, medium and high traffic intensities (up to 30,000 vehicles per day), residential areas (<2,000 vehicles per day), backyards and urban park areas (Fig. 1). The area was divided into 35 grid cells each 200 × 200 m whose midpoints represented the

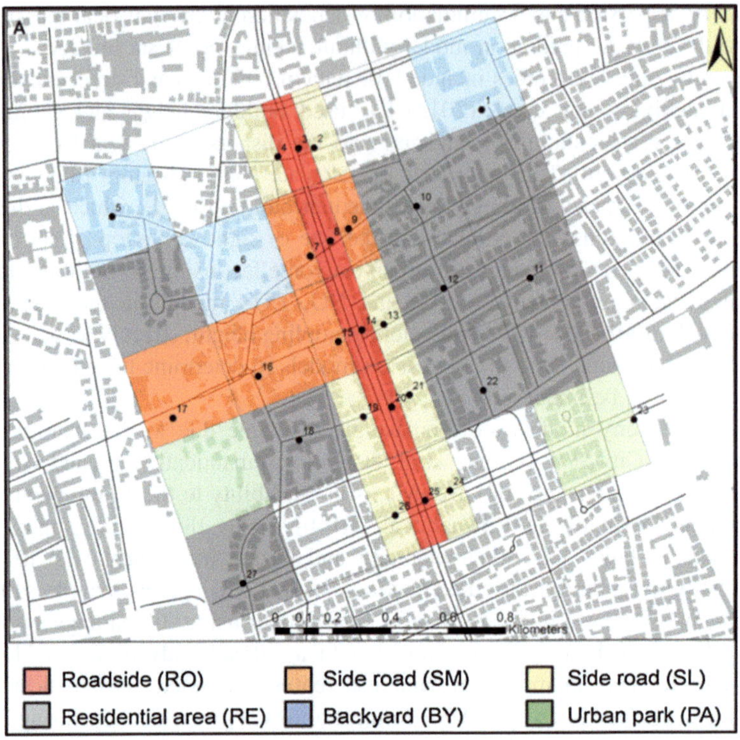


**Fig. 1** Study area in Braunschweig, Germany. It is characterized by street canyons with high traffic intensities, residential areas, backyards and urban park areas (from Ruths et al. 2014)

sampling point at which mobile measurements of particulate air pollutants were conducted (Ruths et al. 2014). The sampling sites were established in six different urban settings, called 'microenvironments'. Microenvironments are defined as micro scale areas with quasi-homogeneous concentrations which can be used to derive characteristic average concentrations (e.g. Ott et al. 2007). At each site, the average concentration of total particle number concentration, the number size distribution and black carbon were measured over 3 min (cf. Sect. 2.2). Measurements were carried out during 8 campaigns in winter (January–March 2013) and 7 campaigns in summer (June–August 2013) during calm weather conditions without rain and low wind speeds <4 $ms^{-1}$.

Monitoring was conducted with a mobile platform consisting of a bicycle connected to a 2-wheel trailer which contained the instruments in a weather protected box. Particle number size distributions were measured with a mobile SMPS the NanoScan scanning mobility particle sizer (TSI 3910, TSI Inc.) which offers a size distribution in the range 11 < Dp <365 nm with a 1 min time resolution. The size fractioned ultrafine number concentration was calculated from the number size distribution as measured by the NanoScan device. Total number concentrations were measured with an Electrical Diffusion Size Classifier (DiSCmini,

MatterAerosol). Black carbon concentrations were measured with an aethalometer AE51 (MageeScientific). To ensure a high accuracy and data quality, all instruments have been synchronized before each measurement campaign.

Furthermore, each campaign was limited to 2 h to reduce a trend of particle concentration due to weather changes over the diurnal course (cf. Ruths et al. 2014 for details).

# 2 Modelling Fine Dust Distribution with Land Use Regression (LUR) Models

## *2.1 Land Use Regression Using 2D Parameters*

Assuming a process related impact on pollutant concentrations, different spatial parameters were grouped into the three process categories to ensure that every process, i.e. emission, dilution and deposition is represented with at least one significant parameter to define the best model (Table 1). These are:

1. Emission category: Spatial parameters belonging to this group can be considered to be characteristic for the emission rate of UFP, i.e. the length and type of a street, reflecting traffic intensity. This category is not significantly affected by seasonal changes;
2. Dilution category: Spatial parameters that define proxies for urban ventilation capability such as the width of streets or the distribution of built up areas; and

**Table 1** List of used spatial parameters and their source next to their extraction method in ArcGIS 10.0

| Category | Sub-categories | Buffer radii (m) | Methods |
|---|---|---|---|
| Emission | Length of different types of streets (residential, primary, tertiary, secondary living) | 50,100 and 200 | Using model builder: generate buffer, intersect with the street, and summarize according the type of the street |
| Dilution | Width of the streets area of the buildings | −100, 200 and 300 | Using model builder: generate buffer, intersect with the buildings, and summarize according the type of the building |
| | Position of the sites (X, Y) | | |
| Deposition | Recreational are (Residential_use500 Green_village500, Commercial500, Construction50, Meadow500, Cemetery500) | 500 | Using model builder: generate buffer, intersect with the buildings, and summarize according the type of the building |

3. Deposition category: Spatial parameters that define potential deposition surfaces such as different types of recreational areas with higher fractions of vegetated surfaces.

Zones of influence, i.e. buffer radii, were specified to reflect the scale of influence appropriate for each process related variable. However, zones representing the influence of a street should be chosen narrow because traffic produced emissions are quite local (buffer radii should be small). In contrast, effects of land use stretch out over a larger area and are more complex as dilution and emission. Thus, larger zones of influence are defined.

Buffers of radii of 50, 100, 200, 300 and 500 m (cf. Table 1) were generated for each measured site using ArcGIS 10.0. Using these buffer radii, all available data (streets, buildings and recreational area) were subsequently clipped to derive the street length, buildings area, and recreational area. Buffer radii of 50, 100 and 200 m were used to derive the length of streets, whereas buffers radii of 100, 200 and 300 m were used to derive the building area. Buffer radii of 500 m were used to derive the area of recreational area.

The algorithm as proposed by Henderson et al. (2007) was used and adopted to build the LUR model for UFP according to the following procedure: all variables were ranked by the absolute strength of their correlation with the measured pollutant. The highest-ranking variable in each sub-category was identified. Afterwards the variables in each sub-category that are correlated (i.e., Pearson's $r \geq 0.6$) with the most highly ranked variables were eliminated to avoid autocorrelation. All remaining variables were implemented into robust linear regression models and variables that were not significant at a 90 % confidence level or that had a coefficient with a counterintuitive sign were rejected. Finally, the last two steps were repeated until convergence.

By following the aforementioned procedures, the most significant variables were selected and implemented in the multi regression model using the open source software R in order to build two models for UFP reflecting the summer and winter campaign, respectively.

## 2.2 Land Use Regression Using 3D Parameters

Here, we enhance the aforementioned LUR model for the prediction of fine dust distribution within a city quarter introducing 3D attributes. Starting from an airborne laser scan derived CityGML-based data base, first, height of the buildings adjacent to the street are selected. Second, the ratio of height and width of the street canyon and, third, volumetric density were used amongst others. However, height was already used in the reference model. The second parameter was described in Tang et al. (2013) and Eeftens et al. (2013). The newly introduced parameter of volumetric density describes the ratio of built volume and air and, therefore, refers to the dilution process. Air volume is calculated as a cylinder with the height of the

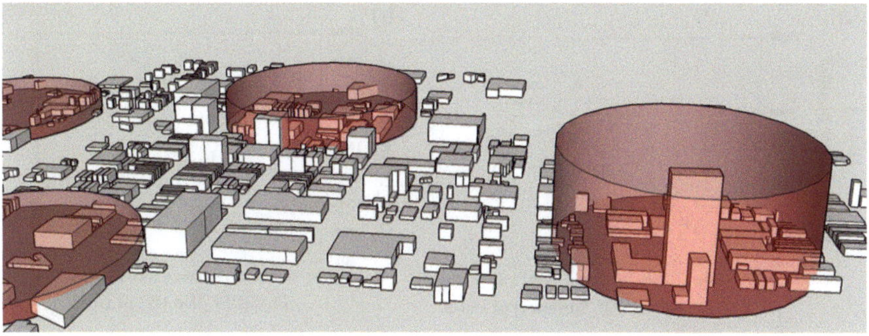

**Fig. 2** Calculation of volumetric density as ratio of built environment and air

highest building lying within a buffer around the point to be modelled (Fig. 2). Buffers of radii of 100 and 200 m were generated for each measured site using ArcGIS 10.0. Using these buffer radii, all the available 3D data were subsequently clipped to derive the buildings volume and to calculate the volumetric density. Then, same aforementioned LUR procedures were used to select the most significant variables and implement them in the multi regression model in order to build two 3D models for UFP, too.

# 3 Results

## *3.1 Evaluation of 2D-LUR and 3D-LUR Models*

Significant 2D variables of each process category (cf. Table 1) were identified and subjected to the multi linear regression process. Then, the best model for the summer and the winter campaign was built.

3D-LUR models used two 3D parameters in each model of summer and winter campaign (volumetric density and width in summer model and volumetric density and the ratio of height and width in winter).

2D-LUR models explained 74 and 85 % of the variance of UFP with root mean square errors (RMSE) of 668 and 1,639 pt. $cm^{-3}$ in summer and winter, respectively.

3D-LUR models explained 41 and 54 % of the variance of UFP with root mean square errors (RMSE) of 868 and 2,541 pt. $cm^{-3}$ in summer and winter, respectively.

A standardized deviation was defined by calculating the ratio of the RMSE to the average measured UFP concentrations across the study area during the summer and winter campaign. The deviation for 2D-LUR accounts to 8 % for summer and 13 % for winter concentrations, 11 % for summer and 19 % for winter concentrations for 3D-LUR, respectively.

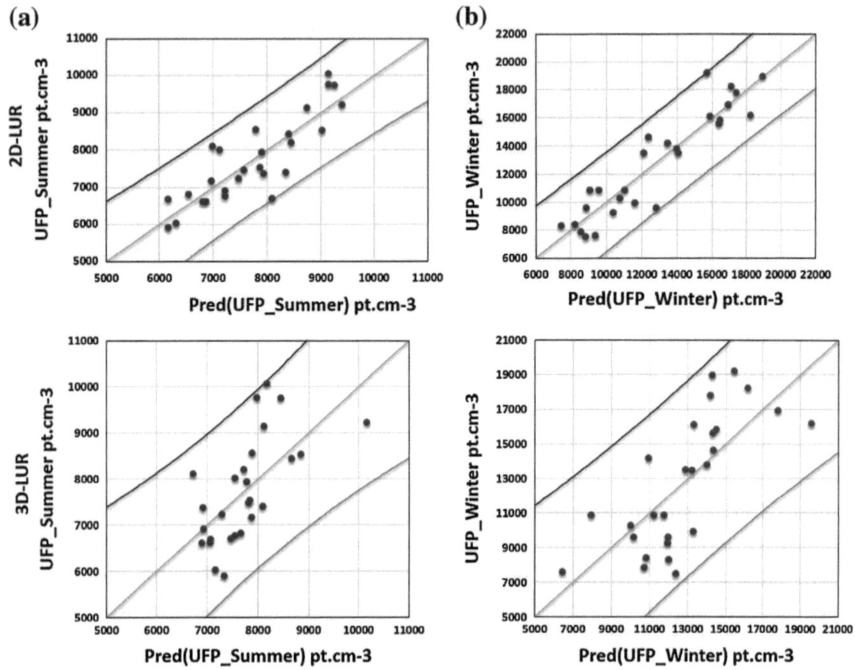


**Fig. 3** Plot of the residuals between the UFP and predicted UFP in **a** the summer and **b** the winter model

Figure 3 depicts that LUR models give evidence for a strong linear trend; no outliers exceeding the confidence interval during summer and winter are observable for both 2D-LUR and 3D-LUR.

The standard residuals of calculated UFP concentration do not exhibit any trend for the summer and winter model (data not shown here).

Leave-one-out cross validation was applied to validate the summer and winter model estimates. The model for each season was rerun on N − 1 data points (N = 27) and then used to predict the excluded measurement. Cross-validation standard error of estimate was calculated for both models. The deviation for 2D-LUR accounts to 9 % for summer and 14 % for winter and is very close to the original model results stating satisfactory model performance.

The deviation for 3D-LUR accounts to 13 % for summer and 22 % for winter. It is evident in Fig. 4 that the different street types are important parameter in the 2D-LUR models. In both seasons the primary road has the highest influence on modelled UFP with relative contribution of 25.86 % during summer and 37.64 % during winter, respectively. This is due to the fact that road traffic acts as a major source for particles in size ranges <100 nm (Morawska et al. 2008), where in 3D-LUR models, volumetric density has the most important influence on the particulate concentration.

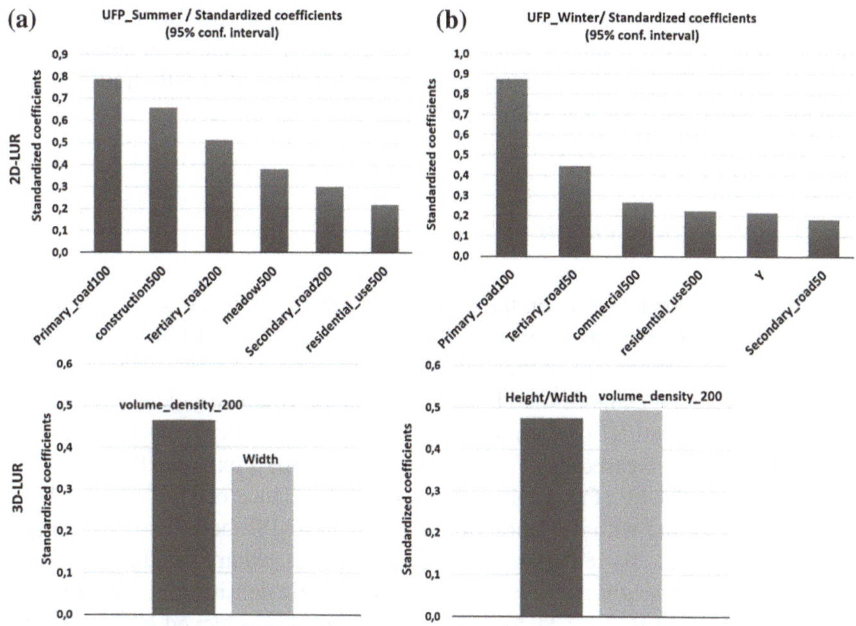


**Fig. 4** Corresponding standardized regression coefficients for **a** the summer and **b** the winter model show the relative influence explanatory variables on the dependent variable, and their significance

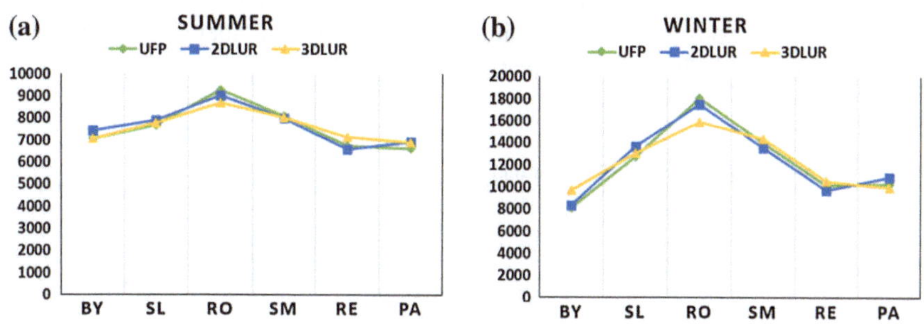


**Fig. 5** The mean variation of UFP concentrations during winter and summer campaign for each microenvironment [measured (UFP), calculated by 2D-LUR and 3DLUR]

Comparison between both models in each microenvironment was conducted (Fig. 5). Generally, both models represented the UFP concentrations very well in all microenvironments.

2D-LUR showed better results than 3D-LUR in BY, RO, and PA for the winter campaign and in BY, RO, and RE for the summer campaign. Both models showed close results in SL and RE for the winter campaign and in SL and SM for the summer campaign.

In spite of the low variance of UFP explained by the 3D-LUR models, the models showed good results comparison with 2D-LUR especially in summer with deviation of 27 % of RMSE between both models. Rather, 3D-LUR showed higher deviation about 38 % of RMSE in winter between both models.

## 3.2 Model Enhancement

The impact of 3D parameters on the model used 2D parameters has been examined by entering the significant parameters of 3D-LUR into the 2D-LUR one by one. It has been noticed that height of the building and the ratio of height and width of the street canyon do not enhance the 2D model, while the volumetric density parameter has slightly increased $R^2$ of the model by 1 %. Furthermore, the average value of the residuals has decreased by 1 %. The new LUR models explained 75 and 86 % of the variance of UFP (Fig. 6) and, therefore, the variance of UFP slightly better than 2D-LUR model with deviation of 2 and 3 % of RMSE between both, summer and winter models. By entering both significant 3D parameters for the winter campaign (ratio of height and width, and the volumetric density) into the 2D-LUR, they have increased $R^2$ by 1 % too. The same result was achieved by entering the two significant 3D parameters for the summer campaign (width of the streets and the volumetric density).

The results indicate that 3D properties have significant impact on the spatial distribution pattern of urban UFP concentrations. We argue that more 3D parameters derived from a CityGML-based 3D city model should enhance the model represented UFP concentration and minimize the required number of 2D parameters needed to predict pollution concentrations quite well.

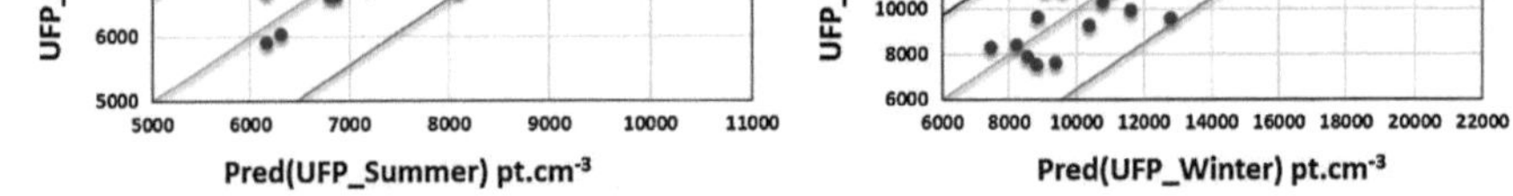


**Fig. 6** Plot of the residuals between the UFP and predicted UFP in **a** the summer and **b** the winter model

# 4 Conclusion

A new modelling approach of using 3D city models available in CityGML has been discussed in this paper to model the UFP distribution that incorporates processes of emission, dilution and deposition of UFP. First results indicate that the 3D parameter enriched model performs quite well in comparison to the 2D model. Although, exhibiting slightly greater deviations, the 3D model needs less parameters than the 2D model. Since ultrafine particle concentration is the result of a complex physical process chain, describing it using simple models with less parameters has many advantages. Fast calculation is a precondition for fast real time calculation and forecast systems.

Entering the significant 3D parameters into the 2D-LUR models has shown a slight enhancement in the final models. In our point of view, this account for the thesis that the local wind field and, therefore, the 3D built environment has a significant influence on the spatial distribution pattern of urban UFP concentrations. However, the research area and the measurement layout described in Sect. “Introduction of Chap. Improving the Consistency of Multi-LOD CityGML Datasets by Removing Redundancy” were initially designed to develop 2D prediction models. That means that it has great variability of properties needed to test significant 2D parameters. Further studies will designate more suitable study areas concerning 3D properties discussed here and, therefore, will yield better results when defining correlations. The research area should have variety concerning 3D properties such as building porosity as an indicator of urban ventilation capacity.

We have shown that combining parameters describing the process of emission and deposition with 3D parameters describing the dilution process and with meteorological parameters can enhance the model. However, further research will focus not only on 3D geometric attributes but on semantic attributes represented in CityGML. This especially accounts for the possibility to incorporate surface properties and window surface proportions of buildings into the model. Since these properties act like a local and temporal sink of small particles, they are of great influence of the fine particle distribution but cannot be stored in a 2D data base.

Semantic attributes available in CityGML may even support the estimation of health situation because of information about the building’s usage and the localisation of its entrance that is more accurate than its address. However, addresses are needed (in addition to pure 3D geometry) because of linking modelling results to epidemiological studies by calculating the distances between residences and major roads and assigning annual fine particulate matter concentrations, derived from the models to each address. Semantics, therefor, may contribute to urban planning.

However, CityGML-based 3D city models still act as incomplete input models for the analysis of particle concentrations. Since no dynamic attributes can be represented in the CityGML2.0 standard, traffic intensities have to be added from other sources. This accounts for the representation of results, as well. Except modelling a specific Application Domain Extension (ADE), there is no way to store different raster based value data back to a CityGML model. Even more seriously, when modelling in different heights using voxel representation model output does not fit into the concept of boundary representation (BRep). The ongoing discussion within the Open Geospatial Consortium clearly should issue the weaving of field based and BRep representation amongst dynamic representation of complex attributes.

**Acknowledgments** The authors would like to thank Matthias Ruths who conducted the mobile measurements of particle and pollutant concentrations.

## References

Arain AM, Blair R, Finkelstein N, Brook RJ, Sahsuvaroglu T, Beckerman B, Zhang L, Jerrett M (2007) The use of the wind fields in a land use regression model to predict air pollution concentrations for health exposure studies. Atmos Environ 41:3453–3464

Brand L, Löwner M-O (2014) Parametrisierung und Identifikation urbaner Straßenkreuzungen im Kontext der Feinstaubmodellierung. "Parameterization and identification of street crossings in the context of fine dust modelling'. Gemeinsame Jahrestagung 2014 der DGfK, der DGPF, der GfGI und des GiN (DGPF Tagungsband 23, 2014)

Brauer M, Hoek G, van Vliet P, Meliefste K, Fischer P, Gehring U, Heinrich J, Cyrys J, Bellander T, Lewne M, Brunekreef B (2003) Estimating long-term average particulate air pollution concentrations: application of traffic indicators and geographic information systems. Epidemiology 14:228–239

Briggs DJ, de Hough C, Gulliver J, Wills J, Elliott P, Kingham S, Smallbone K (2000) A regression-based method for mapping traffic-related air pollution: application and testing in four contrasting urban environments. Sci Total Environ 253:151–167

Brook RD, Rajagopalan S, Pope CA 3rd, Brook JR, Bhatnagar A, Diez-Roux AV, Holguin F, Hong Y, Luepker RV, Mittleman MA, Peters A, Siscovick D, Smith SC, Whitsel L, Kaufman JD (2010) American heart association council on epidemiology and prevention, council on the kidney in cardiovascular disease, and council on nutrition, physical activity and metabolism. particulate matter air pollution and cardiovascular disease: an update to the scientific statement from the American heart association. Circulation. 121:2331–2378

Eeftens M, Beekhuien J, Beelen R, Wang M, Vermeulen R, Brunekreef B, Huss A, Hoek G (2013) Quantifying urban street configuration for improvements in air pollution models. Atmos Environ 72:1–9

Geiser M, Rothen-Rutishauser B, Kapp N, Schürch S, Kreyling W, Schulz H, Semmler M, Im Hof V, Heyder J, Gehr P (2005) Ultrafine particles cross cellular membranes by non-phagocytic mechanisms in lungs and in cultured cells. Environ Health Perspect 113:1555–1560

Gilbert LN, Goldberg SM, Beckerman B, Brook RJ, Jerrett M (2005) Assessing spatial variability of ambient nitrogen dioxide in Montreal, Canada, with a land-use regression model. Air waste Manag Assoc 55:1059–1063

Gröger G, Kolbe TH, Nagel C, Häfele K-H (2012) OGC city geography markup language (CityGML) encoding standard, version 2.0, OGC doc no. 12-019. Open Geospatial Consortium

Henderson S, Beckerman B, Jerrett M, Brauer M (2007) Application of land use regression to estimate long-term concentrations of traffic-related nitrogen oxides and fine particulate matter. Environ Sci Technol. 41(7):2422–8

Hoffmann B, Moebus S, Möhlenkamp S, Stang A, Lehmann N, Dragano N, Schmermund A, Memmesheimer M, Mann K, Erbel R, Jöckel K-H (2006) Residential exposure to traffic is associated with coronary atherosclerosis. Am Heart Assoc Circ 116:489–496

Jerrett M (2011) Spatiotemporal analysis of air pollution and mortality in California based on the American cancer society cohort. Final report

Jerrett M, Beckerman B, Brook J, Finkelstein M, Gilbert N (2006) A land use regression model for predicting ambient concentrations of nitrogen dioxide in Hamilton, Festbrennstoffe, Biokraftstoffe, Biogas

Löwner M-O, Casper E, Becker T, Benner J, Gröger G, Gruber U, Häfele K-H, Kaden R, Schlüter S (2013a) CityGML 2.0—ein internationaler Standard für 3D-Stadtmodelle, Teil 2: CityGML in der Praxis. Zeitschrift für Geodäsie, Geoinformation und Landmanagement, 2, 2013, pp 131–143

Löwner M-O, Benner J, Gröger G, Häfele K-H (2013b) New concepts for structuring 3d city models—an extended level of detail concept for CityGML buildings. In: Murgante B, Misra S, Carlini M, Torre CM, Nguyen H-Q, Taniar D, Apduhan BO, Gervasi O, (eds) ICCSA 2013, Part III, LNCS 7973, Springer, Heidelberg, pp 466–480

Mercer LD, Szpiro AA, Sheppard L, Lindström J, Adar S, Allen R, Avol E, Oron A, Larson T, Liu LJ, Kaufman JD (2011) Comparing universal kriging and land-use regression for predicting concentrations of gaseous oxides of nitrogen ($NO_x$) for the Multi-Ethnic Study of Atherosclerosis and Air Pollution (MESA Air). Atmos Environ pp 4412–4420

Morawska L, Ristovski Z, Jayaratne ER, Keogh DU, Ling X (2008) Ambient nano and ultrafine particles from motor vehicle emissions: characteristics, ambient processing and implications on human exposure. Atmos Environ 42:8113–8138

Ott WR, Steinemann AC, Wallace LA (eds) (2007) Exposure analysis. CRC Press, Taylor & Francis Group, NW, 553 pp

Ruths M, von Bismarck-Osten C, Weber S (2014) Measuring and modelling the local-scale spatio-temporal variation of urban particle number size distributions and black carbon. Atmos Environ

Saraswat A, Apte SJ, Kandlikar M, Brauer M, Henderson BS, Marshall DJ (2013) Spatiotemporal land use regression models of fine, ultrafine, and black carbon particulate matter in New Delhi. Environ Sci Technol 42:12903–12911

Tang R, Blangiardo M, Gulliver J (2013) Using building heights and street configuration to enhance intraurban PM10, $NO_x$, and $NO_2$ Land use regression models. Environ Sci Technol 47:11643–11650

Vardoulakis S, Gonzalez-Flesca N, Fisher BEA (2002) Assessment of traffic-related air pollution in two street canyons in Paris: implications for exposure studies. Atmos Environ 36:1025–1039

Weber S, Kordowski K, Kuttler W (2013) Variability of particle number concentration and particle size dynamics in an urban street canyon under different meteorological conditions. Sci Total Environ 449:102–114

WHO Panel (2013) Review of evidence on health aspects of air pollution—REVIHAAP Project. WHO Regional Office for Europe, Copenhagen, Denmark. http://www.euro.who.int/pubrequest. Accessed on 30, March 2014

# 3D Modelling with National Coverage: Bridging the Gap Between Research and Practice

## Commission II, WG II/2

**Jantien Stoter, Carsten Roensdorf, Rollo Home, Dave Capstick, André Streilein, Tobias Kellenberger, Eric Bayers, Paul Kane, Josef Dorsch, Piotr Woźniak, Gunnar Lysell, Thomas Lithen, Benedicte Bucher, Nicolas Paparoditis and Risto Ilves**

**Abstract** 3D technologies are becoming mature and more and more organisations are investing in 3D models for their areas. For National Mapping and Cadastral Agencies (NMCAs), that have a long history in collating and maintaining countrywide 2D-datasets, the major challenge is how to best adopt a 2D base into 3D environment with an established process of continuous incremental update, and making best use of available resources. To identify best and common practices, as well as remaining research challenges, since 2013 ten NMCAs work together in the

J. Stoter (✉)
Kadaster & TU Delft, Delft, The Netherlands
e-mail: j.e.stoter@tudelft.nl

C. Roensdorf · R. Home · D. Capstick
Ordnance Survey, Southampton, UK

A. Streilein · T. Kellenberger
Swisstopo, Wabern, Switzerland

E. Bayers
IGN, Brussels, Belgium

P. Kane
Ordnance Survey, Dublin, Ireland

J. Dorsch
LDBV, Bavaria, Germany

P. Woźniak
GUGIK, Warszawa, Poland

G. Lysell · T. Lithen
Lantmateriet, Gävle, Sweden

B. Bucher · N. Paparoditis
IGN, Paris, France

R. Ilves
NLS, Helsinki, Finland

© Springer International Publishing Switzerland 2015

M. Breunig et al. (eds.), *3D Geoinformation Science*,
Lecture Notes in Geoinformation and Cartography,
DOI 10.1007/978-3-319-12181-9_13

European Spatial Data Research (EuroSDR) 3D Special Interest Group. This paper reports about an inventory that has been made on the state-of-the art of 3D products and plans that these NMCAs have as well as the challenges they face. The 3D modelling of buildings is explored in more detail, since buildings are prominent features in 3D city and landscape models. In addition, together with road objects (traffic infrastructure), building objects change often and therefore these require efficient update processes.

**Keywords** 3D city and landscape modelling · 3D object reconstruction with national coverage · 3D updating · 3D building modelling

# 1 Introduction

Traditionally the design of georeferencing frameworks, be they geodetic or topographic, has made accommodation for 3D elements. Traditional representation of the third dimension is done through the definition of ellipsoids for geodetic frameworks, the production of reference points and interpolation methods for altitudes, and the representation of height of some features such as buildings, by an attribute. Yet, the encoding of geographical data that constitutes topographical frameworks has always favoured two dimensions because of several reasons. First, the importance of visualisation, on paper or screens, has necessitated the definition of planar projection and representation of the third dimension with 2D features. Second, available technologies for data acquisition were more adapted to 2D. Finally, the methodologies for modelling in 2D, which evolved, have proven to be already very rich for most geo-applications and become entrenched.

The last few years we have witnessed the rapid evolution in 3D visualisation which has put a spotlight on the value of 3D visualisation in raising awareness, particularly of complex phenomena, and promotion, of for example a territory. The growing awareness for our intensively used environment makes 3D information increasingly important, not just in describing the physical environment, but also for modelling abstract features such as flood dynamics, pollution spread and urban analytics in an intuitive and accessible manner.

Besides the many initiatives focusing on generating and using city models, National Mapping and Cadastre Agencies (NMCAs) are also extending their traditional 2D maps into the 3rd dimension. For NMCAs that have a long history in collating and maintaining countrywide 2D-datasets, the major challenge is how to best adopt a 2D base into 3D environment with an established process of continuous incremental update, and making best use of available resources. The ability to retain consistency with the original source is an imperative.

Many NMCAs have made or are making the step towards 3D mapping, although several issues remain on how to generate appropriate 3D data in most effective manner, to the *appropriate* specification, maintain it accordingly, and use it in a wide variety of applications. Ten NMCAs have joined forces in the European Spatial Data Research (EuroSDR) 3D Special Interest Group (3D SIG) to address these issues since they see it as important to further develop in the 3D domain, with help of universities and companies. The NMCAs of the EuroSDR 3D SIG are: GUGIK (Poland), Swisstopo, Ordnance Survey Ireland (OSI), Ordnance Survey UK (OSUK), IGN France, IGN Belgium, Bavarian Agency for Digitisation, High-Speed Internet and Surveying (Germany), Lantmateriet Sweden, NLS Finland and Dutch Kadaster. The purpose of the established EuroSDR 3D SIG is to coordinate the long term research on 3D based on own experiences and developments in both research and industry, by means of describing best practices (e.g. for 3D modelling of buildings), defining a joint research plan, conducting research projects in areas of common interest (e.g. 3D object reconstruction for large areas based on 3D source data that is typically available at NMCAs) and the organization of a workshop series on relevant topics. This paper presents the progresses made in 3D of the ten EuroSDR 3D SIG members. It updates and extends the paper that analysed five of these NMCAs in 2013 (see Stoter et al. 2013a).

The paper provides insights into how results from research are adopted in practice and what further research is required to be able to map complete countries in 3D as part of mainstream spatial information flows. The paper is based on an inventory, the compilation of which is presented first in the methodology. After listing the main research challenges for several aspects of 3D, ranging from data collection, modelling, maintenance to dissemination, specific attention is paid to the approach and issues for modelling and updating 3D building features. Buildings are prominent features in 3D city and landscape models and these features change the most (together with traffic infrastructure). The paper ends with conclusions.

## 2 Methodology

To see the progress and plans of the ten NMCAs, an inventory was made in February 2014. The aim of this inventory was to identify areas of commonality and divergence of 3D approaches of NMCAs. The inventory consisted of two parts. The first part is product-oriented and meant to identify existing and planned 3D products, including the 'issues' associated with getting there. It also focused on current or planned 3D research that is directly related to a product development.

The second part focused on collecting issues that cross 3D products. From both inventories a generic research agenda was defined from the programs that the NMCAs have. This research agenda is currently being further developed after having been discussed at meetings with the NMCAs in March and June 2014.

# 3 Existing and Planned 3D Models

This section presents an analysis of the outcomes of the first part of the inventory, i.e. on existing and planned 3D models.

The available countrywide 3D data produced by NMCAs can be divided into two main products: field data (Digital Terrain Models, Digital Elevation Models, Digital Surface Models, point clouds) and object-oriented data, i.e. topographic features. This latter data can be both 3D (mostly buildings) and 2.5D (surfaces for features like terrain, land use, vegetation, water, road).

Field data is produced by all concerned NMCAs using various methods; see Fig. 1 for the DTM example of OSUK. Storage and dissemination is also done in various formats, such as LAS files (in DBMS), GeoTIFF, TIN, ESRI ASCII Grid, ESRI File Geodatabase 10.1, ASCII X,Y,Z single space, GML (contours).

Field data can be considered as raw data. In contrast, object oriented data can be considered as interpreted data. Several NMCAs produce such 3D object data:

**The Regional Mapping and Cadastre Agency of Bavaria** produces LOD1 and LOD2 building models for their 8.1 million buildings, see Fig. 2. The bases of these buildings consist of cadaster building outlines. LOD1 buildings are generated automatically from Lidar data and updated twice a year with new cadaster data and standard heights derived from number of floors where the Lidar data are not updated. LOD1 buildings are available for all 8.1 million buildings (since 2010). A German-wide LOD1 model will be offered in 2015 via the Working Committee

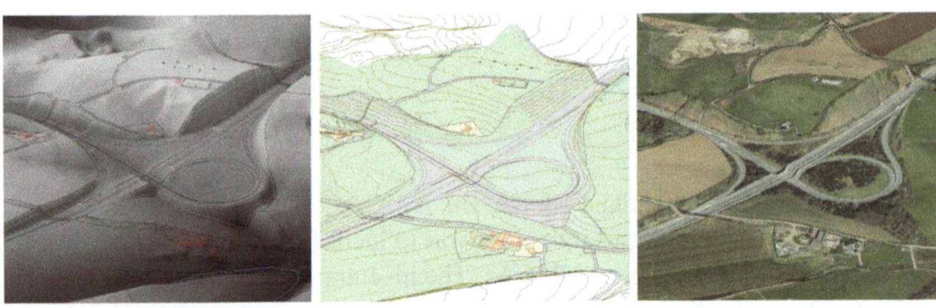

**Fig. 1** DTM produced by OSUK

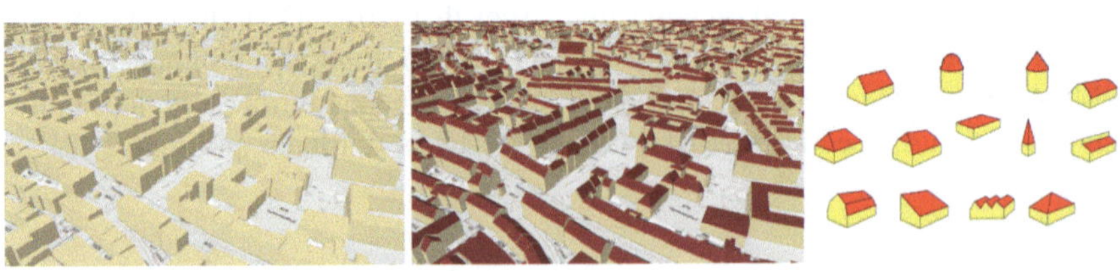

**Fig. 2** Buildings from Bavaria: LOD1 (*left*); LOD2 (*middle*); 13 standard roof shapes from ALKIS®-object catalog for LOD2 buildings (*right*)

of the Surveying Authorities of the States of the Federal Republic of Germany (AdV) in a common CityGML standard.

The LOD2 buildings with roof shapes are also generated from Lidar data using an object catalog with 13 different roof shapes (see Fig. 2, right). Image matching point cloud is used (4 points/$m^2$) for missing buildings in Lidar data. LOD2 buildings are available for 25 % of the buildings and are expected to be complete in 2016. The LOD2 shapes are created with semi-automatic processing and interactive post-processing using the LOD2 reconstruction software "BuildingReconstruction" (company virtualcitySYSTEMS in Berlin).

The update process of LOD2 is being developed using a modified version of the LOD2 reconstruction software. Basis for update information is the terrestrial measurement of buildings by cadaster offices. The semi-automatic update process will start in 2015.

***Storage***

Storage format is CityGML and the buildings are stored in the 3D CityDB implementation of the Technical University of Berlin (Oracle).

***Distribution***

LOD1 buildings are distributed in formats CityGML, KML/KMZ and Shape (with attribute height). German-wide distribution of LOD1 buildings is done by the ZSHH in Cologne (central office for house coordinates and outlines). The LOD2 buildings are distributed in formats CityGML, KML/KMZ, 3D Shape, dxf, and 3ds. The use of a web service for distribution (WFS2.0) for LOD2 is being studied.

**IGN, Belgium** maintains its classical midscale product Inventaire Topo Géographique—Topo Geografische Inventaris (ITGI) in 2.5D: i.e. every x, y-coordinate contains a z-coordinate derived from the 20 m DTM (Fig. 3). The data is created with stereo-restitution, field survey, ortho-interpretation and image correlation (DTM). It was first acquired from 1992 to 2008 and is being updating since 2009.

**Fig. 3** The 3D topographical data set of IGN, Belgium

The data is interactively edited with various software: Intergraph geomedia + ISSG, ArcGIS, Inpho (Match-AT, Match-T), DTM Master, FME, Radius Studio.

***Storage*** is done in Oracle/ESRI Enterprise geodb (SDE).

***Distributed*** via:

- Top10 vector (derived standard product)
- Testbed WMS services (see www.ngi.be)
- Cartoweb WMTS service (see www.ngi.be)

**Other NMCAs**

The 3D data products of swisstopo, Kadaster, IGN France and Ordnance Survey were described in detail (Stoter et al. 2013a). In summary, **swisstopo** maintains their large-scale data in 3D since 2010, while the 2D maps are derived from the 3D swissTLM$^{3D}$ database (O'Sullivan et al. 2008). **The Netherlands** has established a national 3D standard extending CityGML for large-scale data (to be captured at the municipal level), see Stoter et al. (2013b). In addition, the **Netherlands Kadaster** is currently creating a 3D version of their 1:10 k topographic database, using high resolution Lidar data and 2D topography (Oude Elberink et al. 2013). Several possible 3D products are in research at **Ordnance Survey UK (OSUK)**, these are listed in the next table:

| | Datasets |
|---|---|
| New elevation models | DTMs (complete) |
| | DSMs |
| Ground heighting of existing 2D datasets | Topographic features |
| | Road network |
| | Hydrology network (complete) |
| | Rail network (complete) |
| | Tracks and paths network |
| | Address points |
| Additional capture for elevated features | Building heights (LOD1) |
| | 3D buildings and structures (LOD2) |
| | Vegetation |

In addition, in 2014, OSUK released a new product called Building Height Attribute, which attributes 25 million buildings in OS Mastermap with height information (update released in June 2014). The heights are calculated by a fully automatic process that uses remotely sensed data, including aerial imagery and digital surface and terrain models (the same that are used to produce the OS Terrain 5 and 50 products). Attribution is provided for the ground level, the base of the roof (which can be thought of as being similar to eave height) and the highest part of the roof for each building.

**Fig. 4** Marseille old downtown Bati3D

**IGN France** launched the BD Topo product in 1994. This product provides a LOD1 model of buildings as part of a 2.5D geographic information framework, the Référentiel à grande échelle (RGE, for large scale reference framework). The nationwide coverage was completed in 2007. In late 2007 IGN-F launched an innovation project which industrialized research results pertaining to 3D data automated extraction, and chose to primarily rely upon CityGML and a pivot format in this venture. This innovation project was completed by the end of 2009, and resulted in an automated workflow currently used for two production activities:

- Creation of high resolution DTMs from dedicated aerial coverages in rural areas (25 cm ground pixel size, typically 4,000–10,000 km$^2$)
- Creation of 3D city models (LOD 2 level) with texturation extracted from images, from higher resolution and denser aerial coverages, called Bati3D product (Fig. 4).

More recently, IGN-F has launched a project to define national specifications for 3D product, together with users, so that 3D data produced at local level according to these specifications can be integrated in a nationwide 3D referencing framework. These specifications are defined considering the state of the art 3D modelling techniques and standards as well as user expectations. IGN-F has gained experience about user expectations with respect to 3D data through the participation of some of its researchers to the COST action TU0801, Semantic cities and 3D models (Cost 2014). In this action, user requirements were elicited between researcher's specialists of data production, cartography and GIS and researchers in different application domains like urbanism, landscape scenery assessment, or serious games.

# 4 Open Research Issues

In the second part of the inventory, the NMCAs were asked to describe the open issues that they face in their 3D product developments. The mentioned challenges can be categorised in eight types and are further detailed in the remainder of this section:

- Open issues regarding field data
- Product enhancement and 3D object reconstruction
- 3D standards and data modelling
- Maintenance, storage and update
- 3D dissemination and portrayal
- Integration, fusion and consistency
- Tooling
- Users and Customers

## *4.1 Open Issues Regarding Field Data*

The challenges that the NMCAs face for collecting and distributing field data are several:

- Managing and documenting the uncertainty during data capture to produce reference data.
- Integrating various products (points clouds, aerial images, terrestrial images, 2D topographic DB)
- Integrating field data with object-oriented data, e.g. breaklines in DTM with roads or rivers in the topographic data set to ensure consistency between both products
- Update of the DTMs, DSMs and DEMs
- Development of the DTM to support the creation of 2.5D and 3D products
- Integration of surface change detection algorithms to improve the manual update process
- Modelling of 3D hyper surfaces as part of the DTM (such as sheers, overhangs, faults, arcs)

## *4.2 Product Enhancement and 3D Object Reconstruction*

NMCAs are seeking better ways to reconstruct (and interpret) 3D representation of objects such as: buildings, networks, vegetation, structures (distinct to buildings, and include features such as tunnels, bridges and piers), hydrography, power lines etc.

The primary focus of the mentioned issues relates mostly to the reconstruction of building features. For example, more building attributes (including roof type attribution) are required as well as accuracy statements. In addition, complex building footprints and large footprints give problems in automated reconstruction processes. A big challenge for buildings remains the question of how to generate roof structures from point-cloud data while keeping consistency with the footprints maintained in the 2D database?

Different production methods may result in different models, i.e. walls projected from the roof edges (typical for airborne techniques) result usually in larger volumes than the walls extruded from true locations of the footprints (common in terrestrial and oblique techniques). None of them violates the LOD2 specification, but they are considerably different from each other (Biljecki et al. 2014).

For the reconstruction of 3D objects represented with surfaces (mostly triangulated surfaces), the following issues are mentioned:

- How to model bridges (i.e. non-equal crossings) and subsurface elements (e.g. tunnels, caves, etc.)?
- How to support vertical faces (triangle) in the TIN (for example to represent curbs)?
- How to model 3D hyper surfaces such as sheers, overhangs, faults, arcs?

For NMCAs that traditionally have to model large areas, a specific question is how to make capturing of 3D objects on a national level affordable? "As automated as possible" might not always be the most efficient solution, i.e. guaranteeing a valid 3D object output for 60 % of the objects could be preferable over a solution that targets a higher percentage but requires higher levels of human verification. Therefore, besides object reconstruction processes, it is important to be able to identify (i.e. validate) which objects have correctly been generated.

## 4.3 3D Standards and Data Modelling

A main issue for 3D data modelling is that common agreements on the content of 3D data are missing. Even within this group of NMCAs active in 3D, there is no common agreement on terms as 3D, 2.5D, LOD, 3D building, 3D model, 3D data etc. nor on how to model 3D topography like building, roads and water.

The establishment of the 3D standard "CityGML" by the Open Geospatial Consortium (OGC) in 2008 was an important step in the standardization of 3D data, 3D features and their use. The standard resulted in increased 3D datasets, in tandem with the range of applications to which 3D is being applied, e.g. solar mapping, noise modelling, cadastre, etc. (OGC 2012).

However, the CityGML standard is meant as a generic standard and allows an implementation freedom. An issue for wide implementation of CityGML is that it does not offer sufficient guidance on how to uniformly and unambiguously implement the standard: while conformance requirements do exist, they do not cover checking the integrity of CityGML geometries. Furthermore, implementation specifications for CityGML hardly exist, with some exceptions such as the Dutch implementation specification for CityGML (Geonovum 2013) and the modelling handbook published by the German 3D Special Interest Group—data quality working group (SIG 3D 2013a, b). In short, CityGML requires further attention to be able to produce consistent and high-quality 3D geo-information encoded in CityGML. This is currently taken up by the CityGML Quality Interoperability Experiment of OGC (2014).

The NMCAs mentioned the following remaining issues for 3D standards:

- Defining a common, international vocabulary for 3D concepts that all NMCAs use, both for geometry and themes and in line with existing standards (INSPIRE, CityGML, national models and approaches).
- Establishing standards for nationwide 3D models in coherence with 2D data models and preferably based on international agreements, considering user needs and considering complementarities with other industries than GI software (typically the BIM industry).
- Defining specifications for complex structures (canopies etc.) and object types at different levels of details, other than buildings (e.g. trees).
- Defining more advanced 3D spatial data models (vector, voxel, CSG, tetrahedrons, curved surfaces, hyper surfaces etc.)
- Reporting of LOD1 and LOD2 buildings for INSPIRE-monitoring of Annex III theme "Buildings"
- Solving the technical problems with respect to the implementation of the CityGML standard, e.g. current industrial solutions limit flexibility.

Also concerning the foundation principle of CityGML, i.e. Levels of Details, the NMCAs mentioned several open issues:

- How do scale and resolution relate to LOD? A common vocabulary on scale and LOD is missing (beyond the existing conceptual and global vocabulary).
- How to derive 2D from 3D data? How to derive LOD(n − 1) from LODn?

## 4.4 Maintenance, Storage and Update

Maintenance, storage and updating is important for NMCAs to make 3D data useful and reliable.

Database Management Systems (DBMSs) nowadays support 3D geometries. Still the NMCAs mention a need for improved storage and maintenance models and solutions for 3D geodata, i.e. vector, voxel, CSG, tetrahedrons, curved surfaces, hyper surfaces, network, vertical surfaces in a TIN etc. In addition, the consistency with 2D data was also mentioned with respect to storing 3D data. Finally, the storage of massive constraint TINs (i.e. the TriangluatedSurface of CityGML) in coherence with the breaklines (i.e. 2D topography) requires further research.

Several issues were mentioned by the NMCA concerning updating such as efficient updating processes, updating via heights points obtained from dense image matching, and change detection in 3D (see also the next section).

## 4.5 3D Dissemination and Portrayal

Nationwide modelling is traditionally the focus for NMCAs. How can the resulting Big 3D data products be distributed via web services? Are OGC Web Feature Services appropriate? Can the OGC draft specification "Web 3D Service (OGC 2010)" be used? How can 3D data be integrated in existing Web viewers? What are the best formats for the dissemination of 3D data as products?

## 4.6 Integration, Fusion and Consistency

Fusing 3D data brings specific challenges and many issues are still unsolved. The NMCAs mentioned the following open questions:

- What are the main consistency issues in data fusion?
- How to fuse 3D data with other 3D data and how to fuse 3D data with 2D data, i.e. buildings, elevation models, roads, water, vegetation, 3D vector and raster/points?
- How to keep 3D data consistent with 2D data and vice versa?
- How to keep 3D vector data and 2.5D DTM data consistent and 3D vector data with the underlying raw data?
- How to maintain consistency between 3D objects (buildings, roads, bridges, water, vegetation) and terrain?
- What technologies allow automatic establishment of consistency between different data sets?
- Some topics define the landform (e.g. water, roads, bridges, tunnels). How is their relation with a DTM best maintained?
- How to combine terrestrial (cadastral) building footprints with roofs from other sources?
- How to integrate near subsurface buildings?
- How to integrate above and below surface data?
- How to integrate 3D geo-data with 3D data from other domains (Building Information Models, Gaming, 3D Land Administration Model)

## 4.7 Tooling

Tooling for 3D data is required to be able to process the data. What are the frameworks for representing 3D spatial relationships, for 3D spatial analysis and simulation and for 3D navigation and interpolation algorithms?

## 4.8 Uses and Customers

When making the step from 2D to 3D, an important question for NMCAs is what is the business case for 3D, i.e. what are the benefits for an NMCA that justify the investment in 3D? Regarding the value of 3D data, several issues for NMCAs exist:

- What are the Best Practices on added value of 3D above 2D?
- What is the best pricing policy for 3D data (for example open data?)
- How to strengthen the awareness of 3D data? Several NMCAs identified that their 3D data is underused.
- How to determine the economic value of 3D (the value to the economy, or public good)? Thus, what business opportunities can be realised through the application of 3D data?
- What are the required 3D data content, formats and quality?
- For 2.5D and 3D data a related question is "what is the specific task of NMCAs to produce such data and what should be left to the market?" Should an NMCA provide a 3D model of the complete country or produce 3D data on demand? What LOD's should an NMCA provide: basic LOD or more (multi-LOD, texture, UUID/linked data)? Should an NMCA provide photorealistic products in 3D, (as today provision of orthos in 2D) which can range from a 3D-point cloud (+classification + RGB) and a 3D-image model (DSM + texture) to 2.5D/3D-GIS map (DTM + volume + texture).

# 5 Focus on Buildings

The previous section listed the main 3D issues for the participating NMCAs. This section elaborates on one of them, which is the modelling and updating of 3D building features. These features are most prominent features in a 3D city and landscape models and require the most effort to model in 3D. In addition, these features are most frequently updated (together with traffic infrastructure) and therefore buildings are traditionally important objects within topographic data sets that NMCAs produce.

For a selection of the involved NMCAs the modelling approach for buildings are explained. At the end of this section, research issues for maintenance and update of building data are detailed.

## 5.1 Building Modelling Approaches

**Switzerland**

swisstopo captures roofs of buildings on the base of aerial stereo imagery (Fig. 5). XYZ (coordinates and height) are stored for every single roof or roof element. For these roofs a library of possible roofs is used (different than in Bavaria). The footprints are automatically or manually deduced, while walls are automatically generated.

Swisstopo does not map all buildings, since the target scale for the data is approximately 1:10 k. Instead, the building (or element) should have minimum dimensions:

- A roof top line is mapped from 1 m height difference
- Minimal diameter of tower-like buildings is 4 m
- Minimal dimension of mapped buildings is 24 $m^2$ area. The minimal length of one side has to be 8 m or larger.
- Minimal dimension of «Gauben» (roof side add-ons) is 8 m

At this moment almost 600,000 buildings have been captured of the 3 million buildings to be mapped. The mapping speed of an employee is one building per 2 min (including all attributes).

**Bavaria**

The way of building modelling in Bavaria was already explained in the second section [see also Arlinger et al. (2013) and Bavaria (2014)]. Interestingly, in the 2D process important height information is already captured, see Fig. 6. The 2.5D information for footprints in the cadaster enables an update process of 3D buildings via terrestrial field survey.

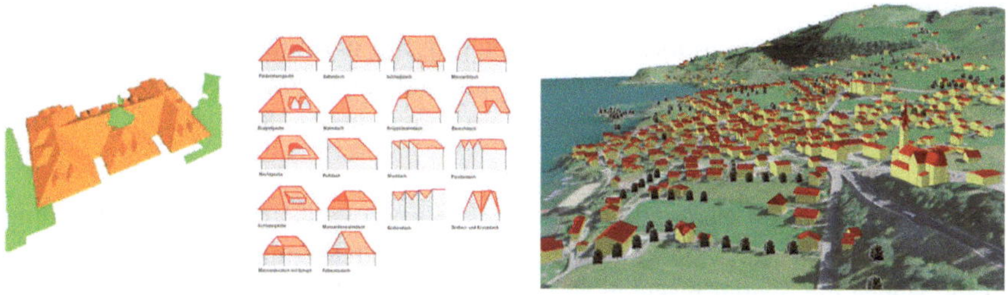

**Fig. 5** Mapping of 3D buildings by swisstopo

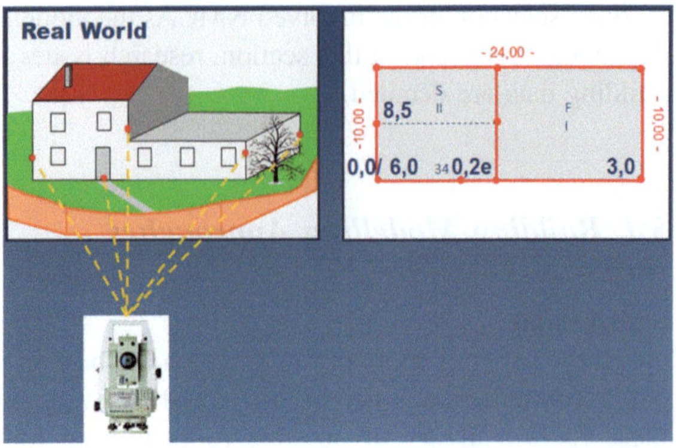


**Fig. 6** Capturing 3D characteristics in 2D data collection process in Bavaria

The automation of LOD2 3D modelling is high (50–85 %). Even though each building model is interactively checked, the output per day per employee is 200–500 LOD2 building models. Bavaria uses minimum requirements for the 3D characteristics of buildings to be mapped in 3D: for buildings smaller than 13 $m^2$, flat roofs are generated. Minimum height of buildings is 1.8 m. Buildings with missing Lidar data get a standard height, for buildings smaller than 25 $m^2$ the standard height given is 3 m and buildings larger than 25 $m^2$ get a standard height of 9 m. In the next years these standard heights will be replaced by roofs from image matching point clouds.

**Sweden**
Lantmateriet, Sweden collects buildings in-house for rural areas and in cooperation with municipalities for urban areas. The buildings are measured in-house in 2.5D with photogrammetry and roof edges are stored in a 2.5D database, see Fig. 7a. The municipalities use photogrammetry or terrestrial measurements to collect in different models. Some municipalities have adopted CityGML or the Norwegian standard SOSI/FKB for photogrammatry, see Fig. 7b. The Norwegian standard supports four levels of detail of buildings depending on user's needs. 3D models can be generated on demand using 2.5D building representations and the DTM. FME and manual labour or data services are used for transfer municipalities' data to Lantmateriet. In Sweden about eight million buildings need to be mapped over an area of 450,000 $km^2$.

**Belgium**
IGN, Belgium captures roofs from stereography and stores the 2.5D geometries in a DBMS. The buildings are part of the digital landscape model, see second section. The 3D characteristics that are captured are shown in Fig. 8.

**France**
IGN France has been conducting researches to improve the modelling of buildings for more than 15 years. 3D modelling is not a research domain driven by obvious users. Hence the overall research strategy in 3D at IGN-F has also been to acquire and distribute innovative 3D data with a concern for cost, quality and performances

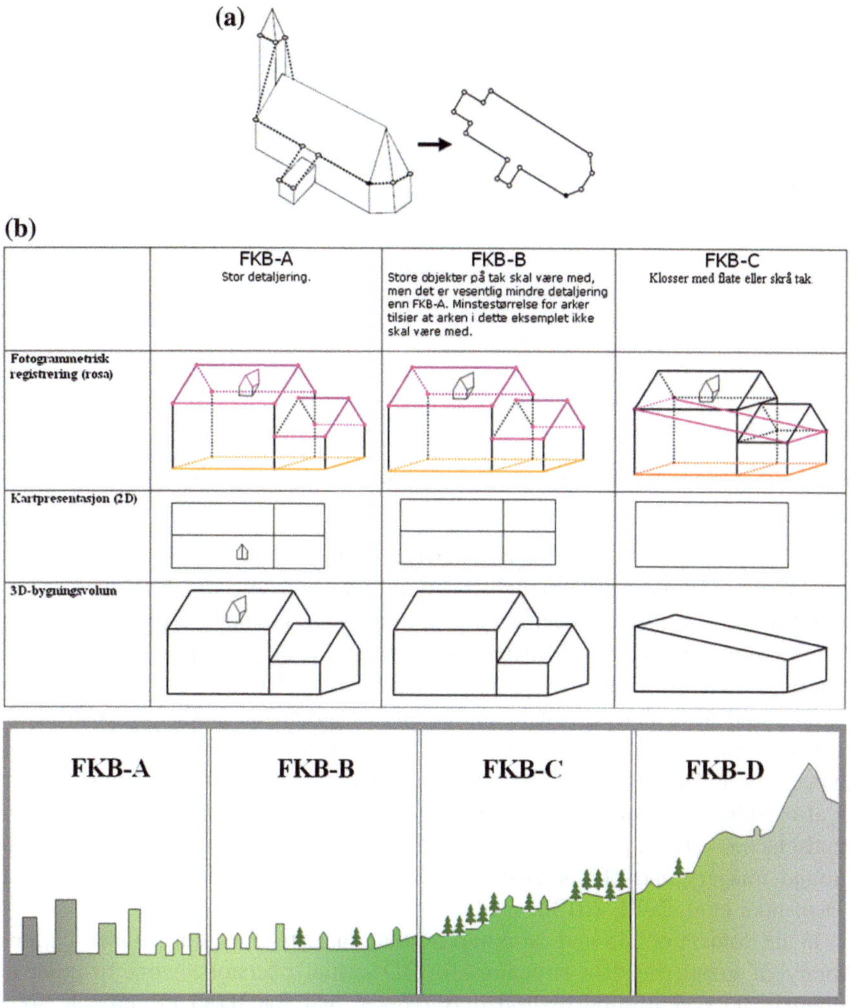


**Fig. 7** **a** Buildings captured in the 2.5D by Lantmateriet, Sweden. **b** 3D modelling approach for some municipalities in Sweden; adopted from Norwegian standard SOSI/FKB

(research led at MATIS lab and LOEMI lab) and to improve GIS and cartography capacities considering these next generation data (research led at COGIT lab).

In the domain of instrumentation, in the late years, contributions have addressed innovative sensors to capture 3D data: photogrammetric cameras (Souchon et al. 2010) and wireless sensor networks to provide ground truth for researchers simulated models for pollution.

In the domain of remote-sensing and photogrammetry, contributions are a mobile mapping systems integrating optical cameras with lidar sensors to achieve a

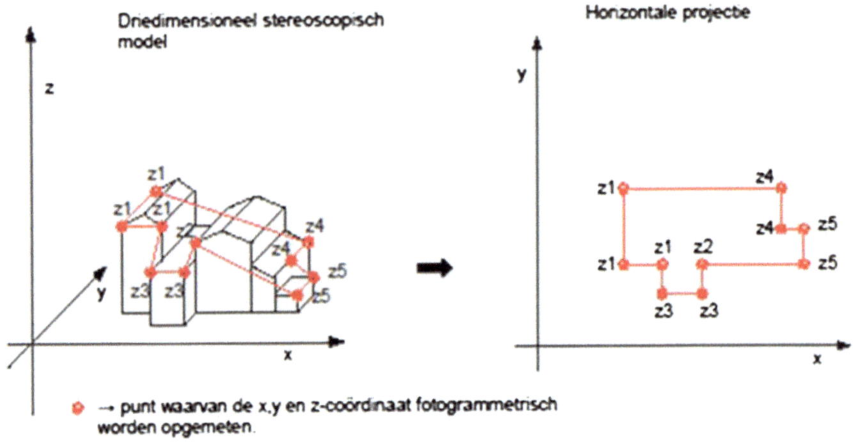


**Fig. 8** 3D characteristics of buildings captured in the 3D modelling process by IGN, Belgium

very high precision (Paparoditis et al. 2012). Chehata et al. (2014) studied sensors specifications for urban 3D model design in order to assess which frequencies are useful in an urban super spectral camera to measure pollution typically. The performance of building reconstruction, edition and quality checking has also been improved (Vallet et al. 2001; Brédif et al. 2013). Current research directions comprehend the enhancement of the quality of automatic modelling with superstructures and cadastral parcels splitting, the qualification of 3D models and the integration of data acquired from mobile mapping systems. 3D model enhancement is also addressed thanks to these mobile mapping systems data. Challenges currently tackled are: registration of laser and 3D model, 3D model texturation with ground imagery and photorealistic 3D façade modelling with deformable grids (Demantké et al. 2012, 2013).

In the domain of GIS and cartography, contributions to building modelling are a library of geometric data structures for 3D spatial operators, work to compare different 3D specification (extruded BDTopo® and Bati3D®) to compute some urban indicators, work with thematic researchers to assess which features, properties and relations have to be derived from the state of the art reference data (3D and 2D and time series) for their application. Brasebin et al. (2011) have proposed an extension of CityGML and specific operators to perform automatic checking of building conformance with urban rules, as well as a methodology to analyse the consequences of urban rules on urban fabric dynamics. Specifically he proposes two models: one for visualisation and one for calculus. Last Bucher et al. (2012) have proposed a vocabulary to describe relations and properties relevant to users of 3D buildings models and to GI modelling experts.

## 5.2 Open Issues for NMCAs to Update and Maintain Building Features

This section explains the specific issues of updating buildings features by NMCAs (which may also be valid for other mapping organisation). The first reconstruction process of a 3D building model (LOD2) is expensive. NMCAs see updating as a unique selling point for the 3D building model produced by NMCAs because regular updates guarantee the long-term maintenance of value and these will avoid a growing inconsistency between 2D and 3D information. Up to 1 % of the buildings are subject to changes every year and existing data sources can be used for the update process.

Three update types for 3D buildings can be distinguished, i.e. a geometric update (building falls away, change of the building, new building), a semantic update (new building function, change in address, roof type name adjusted, number of floors changed, last date of update check change) and updates because of updates in others sources (update of DTM).

Approaches for updating can be a bottom-up approach based on exact 2D footprints. A new building in the 2D database triggers a change in 3D. The 3D geometry can then be acquired via for example terrestrial measurements. Updates in a top-down approach are based on cyclic remote sensing data (new LiDAR data or images for dense image matching). These new data will detect building changes and roofs can be modelled in 3D accordingly. Updating can also be done by a combination of both: new 2D footprints trigger changes of building; intersection of footprints with new 3D data from remote sensing can be used for 3D modelling of single buildings. A significant difference between the two approaches is that the first approach (bottom up) assures consistency between the footprints of 3D buildings and the footprints in the 2D data set, while the second (top down approach) does not.

From the inventory and discussions, we have identified a list of issues for updating 3D building models by NMCAs, which are often also valid for other object types.

These issues are:

- What triggers change in 3D data?
- What is the most efficient updating process (with respect to quality and economy)?
- What requirements should be laid down for the source data?
- What is better given the state-of-the-art: automatic approaches or semi-automatic approaches with post-processing?
- Can heights obtained from dense image matching be used as alternative for Lidar data when updating buildings?
- What are the customer requirements regarding updating, taking into account that change detection with image matching will result in update cycles of 1–3 years and updates via cadastre makes it possible to update in weeks of building modelling in Bavaria was already explainedor months?

  - do users require 100 % consistency between 2D and 3D?
  - do users require the same actuality for 3D data as for 2D data?

- What are the effects of the updating process to the data management?
- How can inconsistencies between 3D data and 2D data be reduced, i.e. in time (different update cycles), in geometry (different scale and data sources) and in semantic information?
- Do we need archiving of 3D buildings and their elements for future 4D applications and how to model these in the database?
- How do all these questions apply to the update of facades and roof textures?

## 6 Conclusion

This paper provides an overview of the 3D modelling activities within a selection of European NMCAs and the challenges they face to build 3D data products within their traditional tasks as geo-information producers. For NMCAs the specific challenge is to map large areas, to assure actuality of the data, to establish a reliable update process and to find a positive business case for 3D data that justify the investments of NMCAs in addition to other organisations that collect 3D data such as municipalities and companies. For one issue, i.e. the modelling and updating of buildings, the current approaches and existing challenges were further detailed.

From this paper it can be concluded that NMCAs have made the step to produce 3D data, while some challenges still need further attention such as maintaining large 3D data sets in DataBase Management Systems; dissemination of the 3D data via web services and via user-friendly formats; automated reconstruction of 3D data (not only buildings); integration of different 3D data and of 2D and 3D data; standardisation in 3D modelling; and, underuse of the 3D data.

## References

Arlinger K, Roschlaub R (2013) Calculation and update of a 3D building model of Bavaria using LiDAR, image matching and cadastre information. In Proceedings of the 8th international 3D GeoInfo conference, Istanbul, Turkey, 28–29 Nov 2013. ISPRS annals of the photogrammetry, remote sensing and spatial information sciences, commission II, WG II/2

Bavaria (2014) http://vermessung.bayern.de, http://www.vermessung.bayern.de/geobasis_lvg/3DGebaeude.html

Biljecki F, Ledoux H, Stoter J (2014) Height references of CityGML LOD1 buildings and their influence on applications. In: Proceedings of the ISPRS 9th 3D GeoInfo conference, Dubai, UAE

Brasebin M, Perret J, Haëck C (2011) Towards a 3D geographic information system for the exploration of urban rules: application to the French local urban planning schemes. In: 28th Urban data management symposium UDMS'11, Delft, Holland, 28–30 Sept. Published in urban and regional data management: UDMS annual 2011, Taylor and Francis, pp 37–50

Brédif M, Tournaire O, Vallet B, Champion N (2013) Extracting polygonal building footprints from digital surface models: a fully-automatic global optimization framework. ISPRS J Photogram Remote Sens 77:57–65

Bucher B, Falquet G, Clementini E, Sester M (2012) Towards a typology of spatial relations and properties for urban applications. In: Conference of 3u3d2012: usage, usability, and utility of 3D city models, Nantes, France, 29–31 Oct 2012. http://dx.doi.org/10.1051/3u3d/201202010

Chehata N, Le_Bris A, Najjar S (2014) Contribution of band selection and fusion for hyperspectral classification. In: Proceedings of the IEEE WHISPERS'14, Lausanne, Switzerland, cityDB. http://www.3dcitydb.net/3dcitydb/welcome/

Cost (2014) Semantic enrichment of 3D city models for sustainable urban development. http://www.cost.eu/domains_actions/tud/Actions/TU0801

Demantké J, Vallet B, Paparoditis N (2012) Streamed vertical rectangle detection in terrestrial laser scans for facade database production. In: XXII congress of the international society of photogrammetry and remote sensing, Melbourne, Australia

Demantké J, Vallet B, Paparoditis N (2013) Facade Reconstruction with Generalized 2.5d Grids. ISPRS Ann Photogramm Remote Sens Spatial Inf Sci II-5/W2:67-72. doi: 10.5194/isprsannals-II-5-W2-67-2013

Elberink SO, Stoter J, Ledoux H, Commandeur T (2013) Generation and dissemination of a national virtual 3D city and landscape model for the Netherlands. Photogram Eng Remote Sens 79:147–158

Geonovum (2013) Technical specifications for the reconstruction of 3D IMGeo CityGML data. http://www.geonovum.nl/sites/default/files/3DFinalReport_2013_1.01_0.pdf

OGC (2010) Draft for candidate OpenGIS® web 3D service interface standard. http://portal.opengeospatial.org/files/?artifact_id=36390

OGC (2012) City geography markup language (CityGML) encoding standard, version 2.0

OGC (2014) The OGC calls for participation in a CityGML data quality interoperability experiment. http://www.opengeospatial.org/pressroom/pressreleases/2054

O'Sullivan L, Bovet S, Streilein A (2008) TLM—the swiss 3D topographic landscape model. In: ISPRS Proceeding, vol. XXXVII. Part B4, pp 1715–1719

Paparoditis N, Papelard J-P, Cannelle B, Devaux A, Soheilian B, David N, Houzay E (2012) Stereopolis II: a multi-purpose and multi-sensor 3D mobile mapping system for street visualisation and 3D metrology. Revue Française de Photogrammétrie et de Télédétection 200: 69–79

SIG 3D (2013a). Teil 1: Grundlagen—Regeln für valide GML Geometrie-Elemente in CityGML. http://wiki.quality.sig3d.org/index.php/Handbuch_fur_die_Modellierung_von_3D_Objekten__Teil_1:_Grundlagen_(Regeln_für_valide_GML_Geometrie-Elemente_in_CityGML)

SIG 3D (2013b) Teil 2: Modellierungshandbuch Gebäude (LOD1, LOD2, LOD3). http://wiki.quality.sig3d.org/index.php/Handbuch_fur_die_Modellierung_von_3D_Objekten_-_Teil_2:_Modellierung_Gebäude_(LOD1,_LOD2_und_LOD3)

Souchon J-P, Thom C, Meynard C, Martin O, Pierrot-Deseilligny M (2010) The IGN CAMv2 system. Photogram Rec 25(132):402–421

Stoter JE, Streilein A, Pla M, Baella B, Capstick D, Home R, Roensdorf C, Lagrange JP (2013a). Approaches of national 3D mapping: research results and standardisation in practice. In Isikdag U (ed) ISPRS annals, vol II-2/W1, s.l. ISPRS, pp 269–278

Stoter JE, van den brink L, Beetz J, Ledoux H, Reuvers M, Klooster R, Janssen P, Penninga F, Vosselman G (2013b) Establishing and implementing a national 3D standard, J Photogram Geoinf (PFG) (in press)

Vallet B, Pierrot-Deseilligny M, Boldo D, Brédif M (2001) Building footprint database improvement for 3D reconstruction: a split and merge approach and its evaluation. ISPRS J Photogram Remote Sens 66(5):732–742

# Out-of-Core Visualization of Classified 3D Point Clouds

**Rico Richter, Sören Discher and Jürgen Döllner**

**Abstract** 3D point clouds represent an essential category of geodata used in a variety of geoinformation applications and systems. We present a novel, interactive out-of-core rendering technique for massive 3D point clouds based on a layered, multi-resolution kd-tree, whereby point-based rendering techniques are selected according to each point's classification (e.g., vegetation, buildings, terrain). The classification-dependent rendering leads to an improved visual representation, enhances recognition of objects within 3D point cloud depictions, and facilitates visual filtering and highlighting. To interactively explore objects, structures, and relations represented by 3D point clouds, our technique provides efficient means for an instantaneous, ad hoc visualization compared to approaches that visualize 3D point clouds by deriving mesh-based 3D models. We have evaluated our approach for massive laser scan datasets of urban areas. The results show the scalability of the technique and how different configurations allow for designing task and domain-specific analysis and inspection tools.

**Keywords** 3D point clouds · LiDAR · Visualization · Point-based rendering

## 1 Introduction

In-situ and remote sensing technology (e.g., airborne, mobile, or terrestrial laser scanning and photogrammetric approaches) allows for efficient and automatic creation of digital representations of spatial environments such as cities and

R. Richter (✉) · S. Discher · J. Döllner
Hasso Plattner Institute, University of Potsdam, Potsdam, Germany
e-mail: rico.richter@hpi.de

S. Discher
e-mail: soeren.discher@hpi.de

J. Döllner
e-mail: juergen.doellner@hpi.de

© Springer International Publishing Switzerland 2015

M. Breunig et al. (eds.), *3D Geoinformation Science*,
Lecture Notes in Geoinformation and Cartography,
DOI 10.1007/978-3-319-12181-9_14

landscapes (Leberl et al. 2010; Lafarge and Mallet 2012). These 3D point clouds are commonly used as an input data for applications, systems, and workflows to derive mesh-based 3D models (Arikan et al. 2013; Beutel et al. 2010) such as for sites, buildings, terrain, and vegetation. These models for example, can be used to create and maintain virtual 3D city models (Lafarge and Mallet 2012; Kolbe 2009), which can be applied in urban planning and development, environmental monitoring, disaster and risk management, and homeland security (Coutinho-Rodrigues et al. 2011). Applications and systems using massive 3D point clouds are faced by increasing availability (e.g., for whole countries), density (e.g., 400 points per $m^2$), and capturing frequency (e.g., once a year). However, they are limited due to their processing strategies that generally do not scale and limited storage capacities. As a remedy they frequently have to reduce the precision and density of the data. To process and analyze large datasets such as massive 3D point clouds out-of-core or external memory algorithms have been designed (Livny et al. 2009; Nebiker et al. 2010; Ganovelli and Scopigno 2012; Rodríguez and Gobbetti 2013). For the inspection and visualization of such datasets out-of-core real-time rendering systems enable an interactive exploration by using specialized spatial data structures and Level-of-Detail (LoD) concepts (Gobbetti and Marton 2004; Wimmer and Scheiblauer 2006; Richter and Döllner 2010; Goswami et al. 2013). These systems generally render all points in a "uniform way" that does not take into account characteristics of different object classes, such as vegetation, building, terrain, street, or water. For example, building façades generally exhibit lower point density in contrast to roofs and terrain. A uniform rendering, therefore, results in gaps between neighboring façade points (Fig. 1), complicating their perception as a continuous surface. If points are rendered by the point primitives of the underlying rendering system (e.g., OpenGL's *GL_POINTS*) they are not scaled according to the camera distance making it difficult to correctly estimate depth differences and leading to visual artifacts due to overlapping of points close to each other. In addition, a uniform rendering does not differentiate between surface characteristics

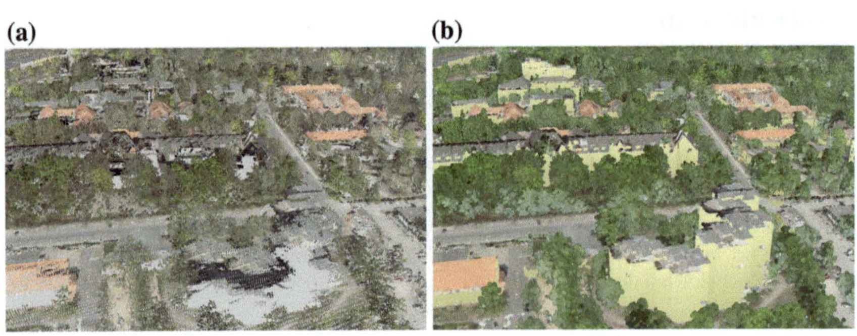


**Fig. 1** **a** Example of a massive 3D point cloud rendered in a uniform way by *GL_POINTS* primitives and textured by aerial photography. **b** Same scene rendered by class-specific point-based techniques: different object classes can be better distinguished, holes on façades are filled, and visual clutter in the background is reduced

such as planar (e.g., terrain), structured (e.g., roof structures), and fuzzy areas (e.g., vegetation), complicating the visual identification and categorization of objects and structures by the user.

We report how the visualization of massive 3D point clouds can be improved based on object class information. Such information are computed with point cloud classification approaches (Lodha et al. 2007; Carlberg et al. 2009; Richter et al. 2013), which typically analyze the 3D point cloud topology, i.e., geometric relationships between points such as connectivity, local flatness, normal distribution, and orientation. We present a novel rendering approach that uses precomputed per-point attributes, such as object class information, color information, and topologic information to adapt the appearance of each point, i.e., its color, size, orientation, and shape. Different photorealistic, non-photorealistic, and solid point-based rendering techniques matching different surface characteristics are selected according to each point's classification. The class-specific rendering techniques can be configured at runtime according to the application and aim of the presentation. To filter and highlight points of specific object classes, focus + context visualization techniques, e.g., interactive and static lenses (Vaaraniemi et al. 2012; Trapp et al. 2008) can be applied. Interactive visualization of massive 3D point clouds, that exceed available memory resources and rendering capabilities, is achieved by storing points in a layered, multi-resolution kd-tree providing an object class specific subdivision of the data.

This paper is structured as follows: Sect. 2 discusses previous work. The system architecture is described in Sect. 3, focusing on point-based rendering techniques and the multi-pass rendering approach. Section 4 introduces the out-of-core rendering visualization based on the layered, multi-resolution kd-tree. In Sect. 5 we evaluate the performance of our system for massive datasets of urban areas. Section 6 gives conclusions and outlines future research directions.

## 2 Related Work

A general overview of point-based rendering is given by Gross and Pfister (2007). Several rendering techniques aim for a photorealistic and, thus, solid visualization of 3D point clouds without holes in the surface (Sibbing et al. 2013; Yu and Turk 2013). These techniques commonly represent points as splats, i.e., oriented flat disks (Botsch et al. 2005; Zwicker et al. 2001), spheres, or particles. To visualize closed surfaces, an adequate size and orientation have to be applied to each point (Kim et al. 2012). These attributes can be calculated in a preprocessing step (Wu and Kobbelt 2004) or on a per-frame basis as proposed by Preiner et al. (2012). However, these techniques are difficult to apply for aerial 3D point clouds because of varying point densities, e.g., on horizontal and vertical structures, as well as on fuzzy and planar areas. In addition, it is difficult to combine these techniques with out-of-core rendering techniques for 3D point clouds because the point density varies depending on the LoD.

Non-photorealistic rendering techniques for 3D point clouds have been proposed by Goesele et al. (2010) and Xu et al. (2004). We extended the silhouette highlighting technique of Xu et al. and added it to our set of rendering techniques. Olson et al. (2011) show how the complete set of silhouette points of a surface can be calculated instant. However, that information comes with the cost of an additional preprocessing step.

Out-of-core rendering systems for 3D point clouds have been presented in Gobbetti and Marton (2004), Wimmer and Scheiblauer (2006), Richter and Döllner (2010), Goswami et al. (2013). These systems use LoD data structures that aggregate or generalize points solely based on spatial attributes. This is not applicable for our purpose because we need to separate points according to their object class at any time during rendering to apply object class specific rendering techniques as well as to render only selected object classes.

Point cloud classification of airborne laser scans has been discussed by several authors in recent years. Identification of building, terrain, and vegetation points is usually achieved by computing and weighting certain features (e.g., normal distribution, surface variation, horizontality) that describe the topology of the local neighborhood of a point (Zhou and Neumann 2008; Lodha et al. 2007). An alternative to that approach is to use attributes specific to the respective scanning technology (Yunfei et al. 2008) (e.g., intensity of returning signals) or information that can be derived from additional geodata covering the same surface area (Kaminsky et al. 2009) (e.g., aerial images, infrastructure maps). In this contribution we compute object class information for each point in a preprocessing pass with a hybrid approach introduced by Richter et al. (2013) that considers topologic features and additional per-point attributes.

In general, object class information is used to extract mesh-based 3D models (Zhou and Neumann 2012) for specific categories such as vegetation, building, or terrain models. However, it is rarely used to enhance the visual quality of a 3D point cloud directly—aside from adapting the colorization of the points. A more advanced rendering approach that does take semantics into account was presented by Gao et al. (2012). They aim for a solid, hole-free visualization of airborne laser scans by resampling terrain segments and by applying a solid rendering style. The purpose of this approach is quite similar to ours. However, our approach supports a larger variety of rendering styles that may be applied to arbitrary object classes at runtime. In addition, the preprocessing in our system is less demanding because we do not differentiate between roof and building points.

## 3 Class-Specific Point-Based Rendering

Our point-based rendering approach uses object class, color, and topologic information on a per-point basis to individualize the appearance of each point. Different point-based techniques are integrated by a multi-pass rendering technique responsible for the final image synthesis.

### 3.1 Data Characteristics

For a given raw 3D point cloud we compute per-point attributes in a preprocessing step. These attributes include the following:

- **Color**. Color or color-infrared values can be extracted from aerial images, ideally captured at the same point in time as the 3D point cloud. These values are generally used for a colorization, e.g., when a photorealistic and natural appearance of the points is required.
- **Object class information**. This attribute denotes to which surface category a point belongs. Typical object classes are vegetation, building, terrain, and water, which can be derived by analyzing the 3D point cloud topology, i.e., local neighborhood of a point. A more detailed subdivision of terrain (e.g., infrastructure, land use) or building points (e.g., commercial, residence) can be made by taking into account additional map data (e.g., infrastructure maps) (Richter et al. 2013).
- **Surface normal**. Per-point normals approximate the surface of the local point proximity. They can be computed efficiently by analyzing the local neighborhood of a point (Mitra and Nguyen 2003) and are used to orientate the point primitive according to the represented surface.
- **Horizontality**. This attribute indicates how vertical the surface normal of a point is oriented, i.e., points representing horizontal surfaces (e.g., flat building roofs) feature higher values than points on vertical surfaces (e.g., building façades) (Zhou and Neumann 2008). The horizontality can be used for a colorization to accentuate detailed object structures (e.g., roof elements).
- **Global height**. This attribute describes the height of a point in relation to all other points that belong to the same object class. Colorizing points based on their global height emphasizes height differences for different objects belonging to the same object class (e.g., trees with different heights).
- **Local height**. The local height describes the height of a point in relation to all points belonging to the same object class in the point's proximity. Using local heights for a colorization allows highlighting edges and differences in the structure of an object (e.g., roof ridges and smokestacks).

All attributes can be used to adapt the appearance of a point, i.e., its color, size, orientation and shape, at run-time. The color of a point can be chosen based on its color value, object class, topology attributes (i.e., surface normal, horizontality, global, or local height), or a combination of these. The orientation of a point can either correspond to its surface normal, the current view direction or a defined uniform vector. In addition, size and shape type of a point can be set dependent on its object class.

### 3.2 Point-Based Rendering Techniques

To efficiently render 3D point clouds, the Graphics Processing Unit (GPU) supports point primitives, such as *GL_POINTS* in OpenGL. However, these primitives have a fixed size in pixels (Shreiner et al. 2013) (e.g., Fig. 1a uses a size of 3 pixel), i.e., their size in object space varies according to their perspective depth. Depending on the view position undersampling, i.e., holes between neighboring points (Fig. 1a—bottom), or oversampling, i.e., visual clutter due to overlapping points (Fig. 1a—top), occurs.

### 3.2.1 Point Splats

To avoid undersampling and oversampling due to changing view positions, the point splats technique renders each point as an opaque disk defined in object space that can be oriented alongside the surface normal (Rusinkiewicz and Levoy 2000; Botsch et al. 2005). The on-screen size depends on the current view position and angle, ensuring a perspective correct visualization (Fig. 2a–f, i). However, the perception of depth differences between overlapping points that are colored homogeneously (e.g., points belonging to the same object class), is generally limited.

### 3.2.2 Point Spheres

We implemented this point-based rendering technique to emphasize the three-dimensional character of a point. The proposed point spheres extend the original splat concept by rendering points as hemispheres instead of flat disks that are always facing the view position and, thus, look like spheres (Rusinkiewicz and Levoy 2000). These hemispheres are created by (1) adding an offset to each depth value of the rendered fragment and by (2) shading each fragment. The depth offset as well as the shading color can be determined by projecting the fragment onto a plane defined by the corresponding splat and by calculating the projected distance of the fragment to the center of the splat. Point spheres are well suited for non-planar and fuzzy surfaces, such as vegetation (Fig. 2g).

### 3.2.3 Silhouette Rendering

Point-based silhouettes highlight and abstract silhouettes and distinctive surface structures (e.g., depth differences). This technique extends the splat rendering approach and was originally proposed by Xu et al. (2004). Similar to the rendering of point spheres, color and depth of each fragment depend on its projected distance to the center of the splat. In addition, the splat is divided into an inner and an outer part. Fragments in the outer part represent the silhouette and are rendered with an increased depth value and a distinct color. As a result, depth discontinuities between overlapping points exceeding a given depth offset are highlighted (Fig. 2h, j, l).

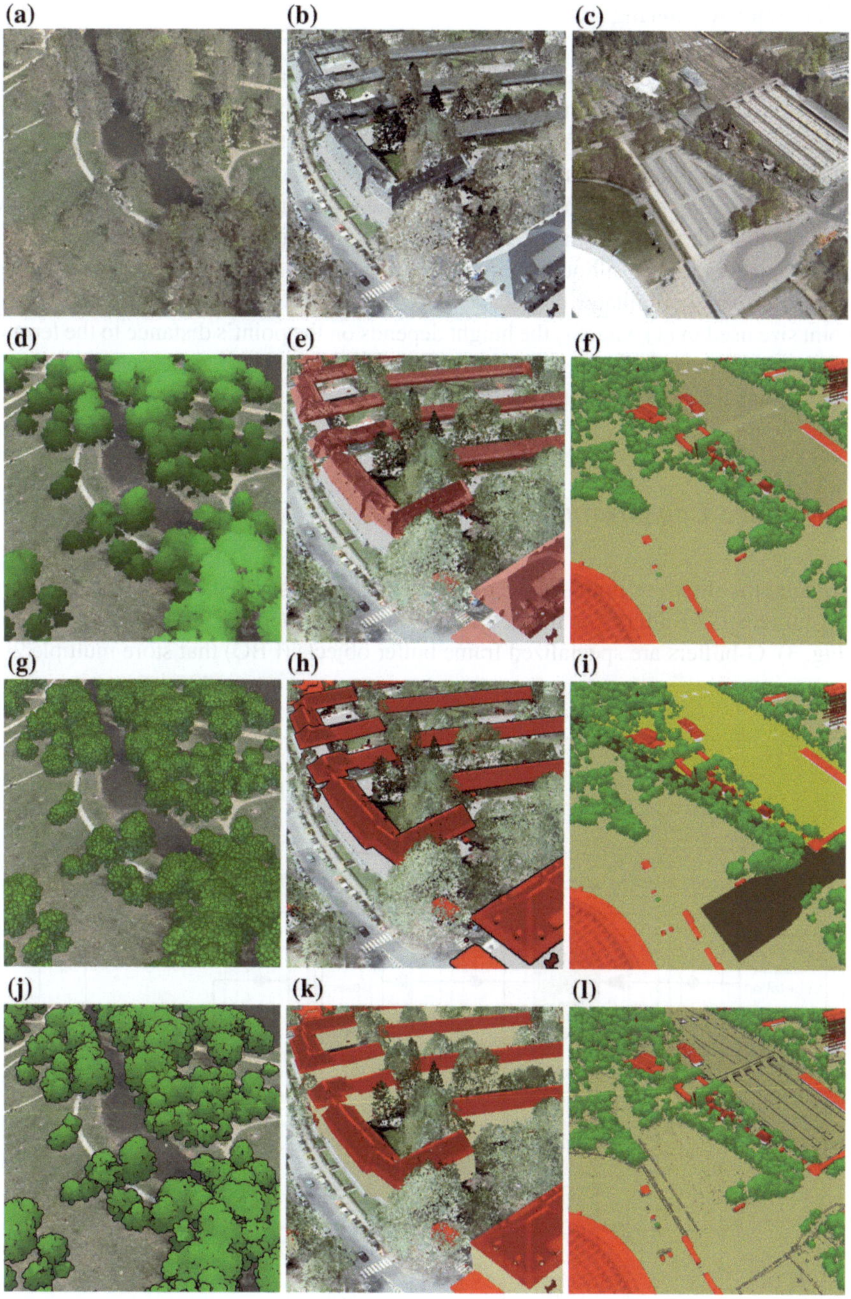


**Fig. 2** Examples of massive 3D point clouds rendered with different rendering setups for vegetation (*left*), buildings (*middle*), and terrain (*right*). **a** Point splats; aerial image colors. **b** Point splats; aerial image colors. **c** Point splats; aerial image colors. **d** Points splats; global height. **e** Point splats; aerial image colors and object class information. **f** Point splats; global height. **g** Point spheres; local height. **h** Silhouette rendering; horizontality. **i** Point splats; object class information. **j** Silhouette rendering; local height. **k** Solid rendering; horizontality. **l** Silhouette rendering; global height

### 3.2.4 Solid Rendering

We developed this point-based rendering technique to render buildings with solid and hole-free façades. As the point density on façades in airborne laser scans is very low in contrast to horizontal structures, the efficient identification of building segments is limited because other structures behind a building are visible through the façade (Gao et al. 2012). To overcome this, we use a second rendering pass to fill the area below roof points with new primitives. The *geometry shader* is used to render (1) a point-based splat, sphere or silhouette equal to the rendering techniques presented above and (2) a quad that imitates the façade below a point. The quad width is equal to the point size used in (1) whereas the height depends on the point's distance to the terrain level. All quads are aligned to the view direction and have the same color or height-based color gradient to create a solid façade look (Fig. 2k).

## *3.3 Image Compositing*

To combine different point-based rendering techniques, we use multi-pass rendering utilizing G-Buffers for image-based compositing (Saito and Takahashi 1990) (Fig. 3). G-buffers are specialized frame buffer objects (FBO) that store multiple 2D

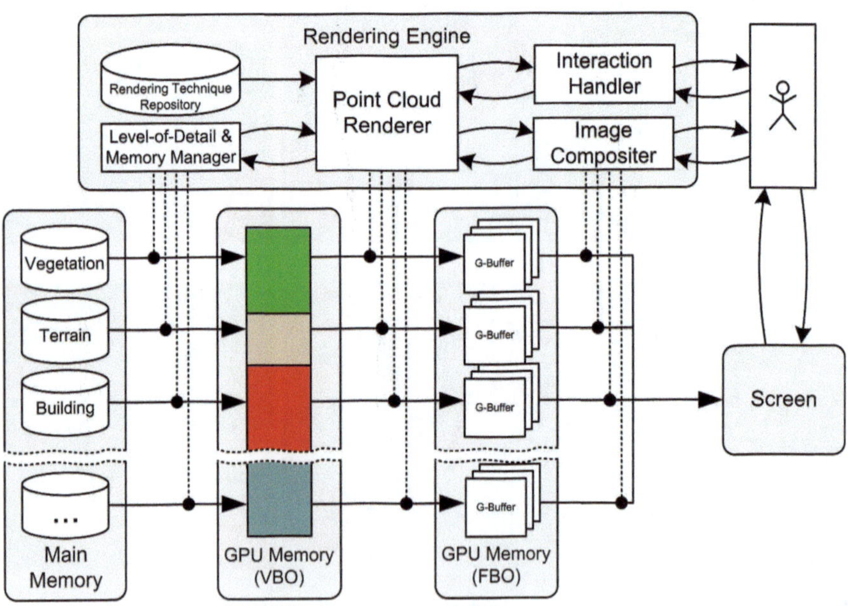


**Fig. 3** Schematic overview of our class-specific point-based rendering system. Categorized by object classes, points are transferred to GPU memory and rendered into separate G-Buffers that are composed to synthesize the final image

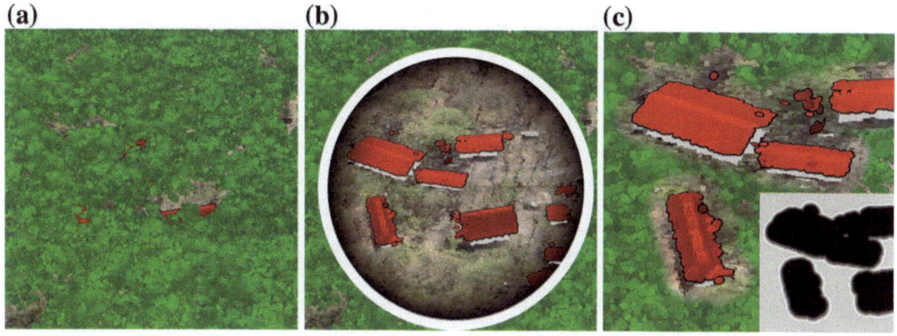


**Fig. 4** Examples of focus + context visualization for classified 3D point clouds. **a** Regular visualization with buildings partially occluded by vegetation. **b** Interactive focus + context lens. **c** Static focus + context lenses positioned around building points

textures for color, depth or normal values. Per object class we have one rendering pass. The results are stored in G-Buffers that are combined by the final rendering pass. This compositing pass allows implementing rendering techniques for focus + context visualization (Vaaraniemi et al. 2012; Trapp et al. 2008) such as interactive lenses (Fig. 4b). Moreover, object class specific visibility masks, i.e., static lenses, can be computed and applied during the rendering to highlight occluded structures (Fig. 4c). Point-based rendering techniques can be independently selected, combined and configured at run-time to adjust the appearance of each object class.

# 4 Out-of-Core Rendering

The interactive visualization of massive 3D point clouds exceeding available memory resources and rendering capabilities and demands for out-of-core rendering techniques that combine LoD concepts, spatial data structures, and external memory algorithms. We developed a *layered, multi-resolution kd-tree* for massive 3D point clouds that have been attributed with object class information. It is characterized by the following properties:

- Object class specific subdivision of the data to enable a selective access and visualization (e.g., only building points).
- Adaptive multi-resolution LoDs to preserve a defined rendering budget (e.g., 30 frames per second).
- Efficient and adaptive memory management (e.g., by using equal-sized LoD chunks).
- Object class specific LoD selection to fulfill different requirements for specific rendering techniques (e.g., varying point densities).

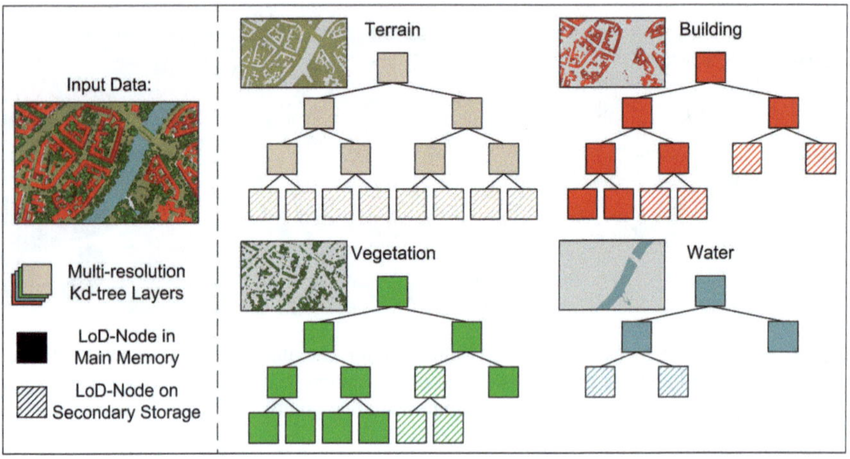


**Fig. 5** Schematic overview showing the structure of our layered, multi-resolution kd-tree. For each object class a separate multi-resolution kd-tree is maintained

## *4.1 Layered Multi-resolution Kd-tree*

Most spatial data structures use kd-tree, quadtree, or octree derivations to arrange 3D point clouds in a preprocessing step (Rusinkiewicz and Levoy 2000; Gobbetti and Marton 2004; Wimmer and Scheiblauer 2006; Richter and Döllner 2010; Goswami et al. 2013). The construction of quadtrees and octrees can be performed faster in contrast to kd-trees because there is no need to sort the points. However, the use of quadtrees and octrees for irregular and sparse distributed data, e.g., airborne laser scans, results in tree nodes with a varying number of points. Out-of-core memory management has to implement efficient caching and memory swapping mechanisms that benefit from equal-sized data chunks. For that reason, we decided to use kd-trees to arrange the data. All points belonging to the same object class are arranged in a sub-tree consisting of nodes with an equal number of points (Fig. 5). Each of these nodes corresponds to a LoD for a spatial area with the root node representing the overall expansion of the 3D point cloud and child nodes subdividing the area of their parent node. Each point is stored only once in the tree, and all nodes together are equal to the input 3D point cloud.

### 4.1.1 Construction

The layered, multi-resolution kd-tree is constructed in a preprocessing step. It can be stored on secondary storage and therefore applied for arbitrary sized 3D point clouds. First, the given 3D point cloud is subdivided based on object classes. Second, for each object class the corresponding points are arranged in a multi-resolution kd-tree.

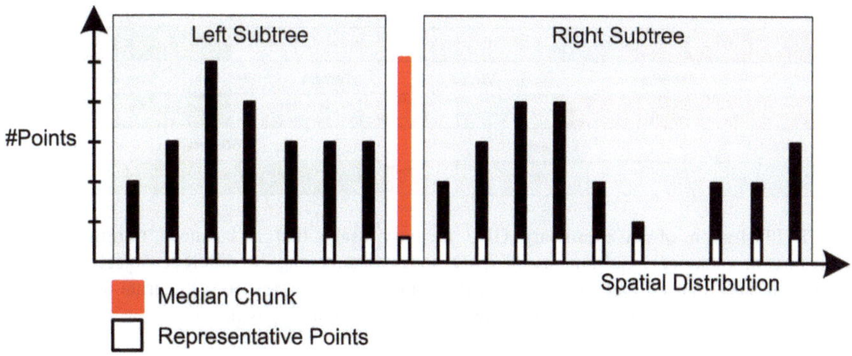


**Fig. 6** Illustration of the histogram-based construction of the kd-tree to reduce preprocessing times for massive 3D point clouds

The construction of a kd-tree with an equal number of points per node, i.e., a *balanced kd-tree*, is implemented by a multi-pass histogram-based approach that avoids a time-consuming sorting of the entire data for each tree level. In a first pass, we iterate over the 3D point cloud to fill a histogram that describes the spatial distribution and extent of the data. Similar to a voxel grid, the histogram organizes points into a number of equal-sized spatial chunks. For each chunk, the number of points belonging to the respective area and a representative point are stored (Fig. 6). Based on the number of points per chunk and the spatial extent of the histogram, a median chunk can be determined that contains the median point required to construct the kd-tree. A second iteration over the 3D point cloud is used to fill up the current node with representative points (i.e., to create a LoD) and to assign all points to the left or right part of the tree. Only points belonging to the median chunk need to be sorted to determine the exact median element. The median element for the split is chosen so that the number of points to the left is a multiple of the number of the points stored per node. This is important to construct a balanced kd-tree with equal sized nodes with exception of one leaf node. The out-of-core construction process subdivides point data on the file system until data chunks can be processed in main memory.

## *4.2 Layered Kd-tree Rendering*

The rendering process can be divided into three stages that are performed per frame. The first stage is responsible for the data provision, caching, and transferring of points from secondary storage to main memory as well as from main memory to GPU memory using the layered, multi-resolution kd-tree. The second stage applies one of our point-based rendering techniques (Sect. 3) to all points belonging to the

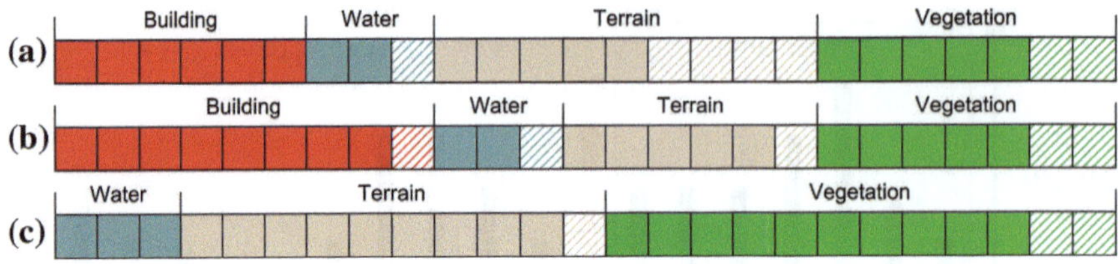


**Fig. 7** Illustration of an exemplary GPU memory usage that is balanced during rendering according to memory requirements of LoD nodes that belong to different object classes. **a**, **b** Illustrate how unused memory is assigned to other object classes. **b**, **c** Illustrate the balancing process when the visualization of one object class (e.g., building) is disabled

respective object class. The last stage seamlessly combines all class-specific rendering results into one final image (Sect. 3.3).

At first, the root nodes of all class-specific sub-trees are loaded into main memory. Each chunk is equal to a LoD node and is mapped into a vertex buffer object (VBO) resident in GPU memory. The VBO is divided into equal sized chunks that can store exactly one LoD node. The layered, multi-resolution kd-tree is used to determine LoD nodes that need to be transferred to or can be removed from the VBO. The decision to add or remove a LoD node from memory depends on the projected node size (PNS). Therefore, the bounding sphere of the node is projected into screen space, and the number of covered pixels is compared to the number of points per node (Richter and Döllner 2010). The threshold applied to the PNS depends on the point-based rendering technique, available memory, and computing capability of the GPU. Each object class has its own memory budget (Fig. 7) and is balanced permanently during the rendering process because the amount of memory required by an object class may vary due to the following reasons:

- Only a small number of points belonging to an object class is visible during the exploration.
- Visualization of certain object classes is disabled.
- Close up views require a high point density for an object class (e.g., for buildings).

Object classes can be rendered with different LoDs because the required number of points for an appropriate rendering result depends on the structure. For example, buildings may require to be rendered with more points due to detailed roof structures in contrast to terrain or vegetation that can be rendered with fewer points. To ensure a hole-free surface, the lower point density can be compensated by using larger primitives, e.g., splats for terrain or spheres for vegetation.

# 5 Results and Applications

We have evaluated the presented system and all implemented point-based rendering techniques with three massive 3D point clouds containing up to 80 billion points (Table 1). For implementation we used C++, OpenGL, GLSL, and OpenSceneGraph. Measurements and tests were performed on an Intel Xeon CPU with 3.20 GHz, 12 GB main memory, and a NVIDIA GeForce GTX 770 with 2 GB device memory.

As shown in Fig. 8, interactive frame rates can be achieved for each rendering technique as long as the overall number of rendered points does not exceed a certain threshold (e.g., 6 million points for the solid rendering approach). The highest frame rate could be observed for *GL_POINTS*, which was expected since these primitives are supported natively by the GPU. Point Spheres as well as our solid and silhouette rendering approach extends the concept of Point Splats and increase the computational effort during rendering. Consequently, lower frame rates were achieved when using these techniques for rendering as opposed to Point Splats. Furthermore, the performance for Point Spheres is higher than for Point Silhouettes due to a more hardware demanding shading implementation (e.g., conditional branching). Since the proposed out-of-core rendering approach limits the number of rendered points by dynamically selecting them, arbitrarily large datasets with varying point densities can be rendered in real-time as well (Table 2).

**Table 1** Characteristics of the datasets used to evaluate the performance of the presented point-based rendering approach

| | Dataset 1 | Dataset 2 | Dataset 3 |
|---|---|---|---|
| Point density (pts/m$^2$) | 10 | 28 | 100 |
| Number points (billion) | 5 | 7.1 | 80 |
| Data size (GB) | 112 | 159 | 1,788 |

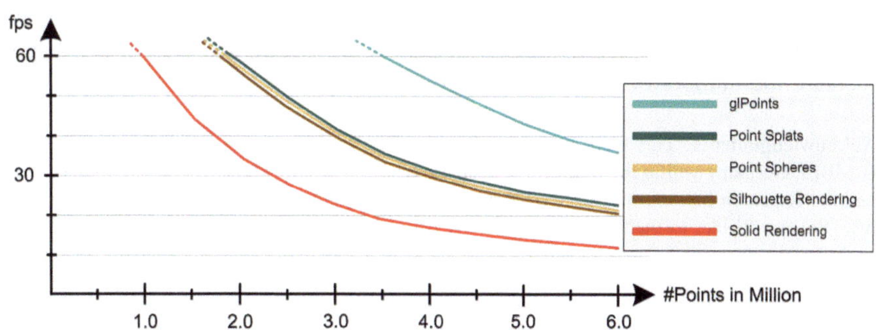


**Fig. 8** Rendering performance in frames per second (fps) using different sized subsets of the datasets from Table 1

**Table 2** Rendering performance in frames per second (fps) using the proposed out-of-core rendering approach. Each dataset is evaluated for a close and a far perspective

| | Dataset 1 | | Dataset 2 | | Dataset 3 | |
|---|---|---|---|---|---|---|
| | Far | Close | Far | Close | Far | Close |
| #Rendered points in million | 2.32 | 0.50 | 3.42 | 0.85 | 4.85 | 1.04 |
| GL_POINTS | 86.39 | 378.07 | 60.02 | 246.12 | 40.24 | 194.35 |
| Point splats | 51.84 | 214.32 | 32.27 | 138.67 | 23.01 | 108.63 |
| Point spheres | 49.57 | 203.81 | 28.31 | 133.72 | 22.35 | 107.07 |
| Silhouette rendering | 46.07 | 195.65 | 26.66 | 127.38 | 22.18 | 106.97 |
| Solid rendering | 27.32 | 100.13 | 20.22 | 63.78 | 18.74 | 59.45 |
| Combination 1 (Fig. 2, row 3) | 40.51 | 200.97 | 27.33 | 128.73 | 22.45 | 107.75 |
| Combination 2 (Fig. 2, row 4) | 33.28 | 126.31 | 22.21 | 80.80 | 19.90 | 68.47 |

# 6 Conclusions and Future Work

We have shown that out-of-core rendering for massive 3D point clouds can be improved by using point-specific attributes such as topologic or semantic information. In particular, object class information can be used to select specialized point-based rendering techniques that take into account class-specific surface characteristics (e.g., solid, planar, non-planar, fuzzy). In addition, it enables focus + context techniques, e.g., lenses for filtering and highlighting. This way we can improve the visual appearance and facilitate recognition of objects within 3D point clouds. Furthermore, our approach offers many degrees of freedom for graphics and interaction design. This approach also allows us to dissolve occlusion and enable a task-specific interactive exploration. The proposed layered, multi-resolution kd-tree enables in addition to a spatial data selection an object class specific selection of LoDs. Hence, memory and processing resources can be used economically and adaptively. In future work, we plan to integrate point-based rendering techniques that enable a per-frame reconstruction of object surfaces (Preiner et al. 2012), e.g., for terrain or roof points. In addition, we want to combine 3D point clouds from aerial scans with data from mobile and terrestrial scans to increase the number of available object classes.

**Acknowledgements** This work was funded by the Federal Ministry of Education and Research (BMBF), Germany within the InnoProfile Transfer research group “4DnD-Vis” (www.4dndvis.de) and the Research School on ‘Service-Oriented Systems Engineering’ of the Hasso Plattner Institute. We would like to thank virtualcitySYSTEMS for providing datasets.

## References


Arikan M, Schwärzler M, Flöry S, Wimmer M, Maierhofer S (2013) O-snap: optimization-based snapping for modeling architecture. ACM Trans Graph 32(1):6:1–6:15

Beutel A, Mølhave T, Agarwal P (2010) Natural neighbor interpolation based grid DEM construction using a GPU. In: 18th SIGSPATIAL international conference on advances in geographic information systems, pp 172–181

Botsch M, Hornung A, Zwicker M, Kobbelt L (2005) High-quality surface splatting on today's GPUs. In: Eurographics symposium on point-based graphics, pp 17–24

Carlberg M, Gao P, Chen G, Zakhor A (2009) Classifying urban landscape in aerial lidar using 3D shape analysis. In: 16th IEEE international conference on image processing, pp 1701–1704

Coutinho-Rodrigues J, Simão A, Antunes C (2011) A GIS-based multicriteria spatial decision support system for planning urban infrastructures. Decis Support Syst 51(3):720–726

Ganovelli F, Scopigno R (2012) OCME: out-of-core mesh editing made practical. Comput Graph Appl 32(3):46–58

Gao Z, Nocera L, Neumann U (2012) Visually-complete aerial LiDAR point cloud rendering. In: 20th international conference on advances in geographic information systems, pp 289–298

Gobbetti E, Marton F (2004) Layered point clouds: a simple and efficient multiresolution structure for distributing and rendering gigantic point-sampled models. Comput Graph 28(6):815–826

Goesele M, Ackermann J, Fuhrmann S, Haubold C, Klowsky R, Steedly D, Szeliski R (2010) Ambient point clouds for view interpolation. ACM Trans Graph 29(4):95:1–95:6

Goswami P, Erol F, Mukhi R, Pajarola R, Gobbetti E (2013) An efficient multiresolution framework for high quality interactive rendering of massive point clouds using multi-way kd-trees. Visual Comput 29(1):69–83

Gross M, Pfister H (2007) Point-based graphics. Morgan Kaufmann Publishers Inc., Los Altos, CA

Kaminsky R, Snavely N, Seitz S, Szeliski R (2009) Alignment of 3D point clouds to overhead images. Computer vision and pattern recognition workshops, pp 63–70

Kim HJ, Öztireli AC, Gross M, Choi SM (2012) Adaptive surface splatting for facial rendering. Comput Animation Virtual Worlds 23(3–4):363–373

Kolbe TH (2009) Representing and exchanging 3D city models with CityGML. In: 3D geo-information sciences, chap 2, pp 15–31. Springer, Berlin

Lafarge F, Mallet C (2012) Creating large-scale city models from 3D-point clouds: a robust approach with hybrid representation. Int J Comput Vision 99(1):69–85

Leberl F, Irschara A, Pock T, Meixner P, Gruber M, Scholz S, Wiechert A (2010) Point clouds: lidar versus 3D vision. Photogram Eng Remote Sens 76(10):1123–1134

Livny Y, Kogan Z, El-Sana J (2009) Seamless patches for GPU-based terrain rendering. Visual Comput 25(3):197–208

Lodha SK, Fitzpatrick DM, Helmbold DP (2007) Aerial lidar data classification using AdaBoost. In: sixth international conference on 3-D digital imaging and modeling (3DIM), pp 435–442

Mitra NJ, Nguyen A (2003) Estimating surface normals in noisy point cloud data. In: 19th annual symposium on computational geometry, pp 322–328

Nebiker S, Bleisch S, Christen M (2010) Rich point clouds in virtual globes—a new paradigm in city modeling? Comput Environ Urban Syst 34(6):508–517

Olson M, Dyer R, Zhang H, Sheffer A (2011) Point set silhouettes via local reconstruction. Comput Graph 35(3):500–509

Preiner R, Jeschke S, Wimmer M (2012) Auto splats: dynamic point cloud visualization on the GPU. In: Proceedings of eurographics symposium on parallel graphics and visualization, pp 139–148

Richter R, Döllner J (2010) Out-of-core real-time visualization of massive 3D point clouds. In: 7th international conference on computer graphics, virtual reality, visualisation and interaction in Africa, pp 121–128

Richter R, Behrens M, Döllner J (2013) Object class segmentation of massive 3D point clouds of urban areas using point cloud topology. Int J Remote Sens 34(23):8408–8424

Rodríguez M, Gobbetti E (2013) Coarse-grained multiresolution structures for mobile exploration of gigantic surface models. In: SIGGRAPH Asia symposium on mobile graphics and interactive applications, pp 4:1–4:6

Rusinkiewicz S, Levoy M (2000) QSplat: a multiresolution point rendering system for large meshes. In: ACM SIGGRAPH, pp 343–352

Saito T, Takahashi T (1990) Comprehensible rendering of 3-D shapes. SIGGRAPH computer Graph 24(4):197–206

Shreiner D, Sellers G, Kessenich JM, Licea-Kane BM (2013) OpenGL programming guide: the official guide to learning OpenGL, version 4.3, 8th edn. Addison-Wesley, Reading

Sibbing D, Sattler T, Leibe B, Kobbelt L (2013) SIFT-realistic rendering. In: International conference on 3D vision, pp 56–63

Trapp M, Glander T, Buchholz H, Döllner J (2008) 3D Generalization lenses for interactive focus + context visualization of virtual city models. In: 12th international conference on information visualisation, pp 356–361

Vaaraniemi M, Freidank M, Westermann R (2012) Enhancing the visibility of labels in 3D navigation maps. Lecture notes in geoinformation and cartography, pp 23–40

Wimmer M, Scheiblauer C (2006) Instant points: fast rendering of unprocessed point clouds. In: Eurographics symposium on point-based graphics, pp 129–137

Wu J, Kobbelt L (2004) Optimized sub-sampling of point sets for surface splatting. Comput Graph Forum 23(3):643–652

Xu H, Nguyen MX, Yuan X, Chen B (2004) Interactive silhouette rendering for point-based models. In: Eurographics symposium on point-based graphics, pp 13–18

Yu J, Turk G (2013) Reconstructing surfaces of particle-based fluids using anisotropic kernels. ACM Trans on Graph 32(1):5:1–5:12

Yunfei B, Guoping L, Chunxiang C, Xiaowen L, Hao Z, Qisheng H, Linyan B, Chaoyi C (2008) Classification of LIDAR point cloud and generation of DTM from LIDAR height and intensity data in forested area. In: International society for photogrammetry and remote sensing congress, pp 313–318

Zhou QY, Neumann U (2008) Fast and extensible building modeling from airborne lidar data. In: 16th ACM SIGSPATIAL international conference on advances in geographic information systems, pp 1–8

Zhou QY, Neumann U (2012) 2.5D building modeling by discovering global regularities. In: Computer vision and pattern recognition, pp 326–333

Zwicker M, Pfister H, van Baar J, Gross MH (2001) Surface splatting. In: ACM SIGGRAPH, pp 371–378

# Modeling Visibility in 3D Space: A Qualitative Frame of Reference

**Paolo Fogliaroni and Eliseo Clementini**

**Abstract** This paper introduces and formalizes a frame of reference for projective relations in 3D space that can be used to model human visual perception. While in 2D space visibility information can be derived from the concept of collinearity (thus, as ternary relations), in 3D space it can be derived from coplanarity, which calls for quaternary relations. Yet, we can retain ternary relations by anchoring our frame to an ubiquitous reference element: a general sense of vertical direction that, on Earth, can be the expression of gravity force or, in other cases, of the asymmetries of an autonomous agent, either human or robotic, that is, its vertical axis. Based on these observations, the presented frame of reference can be used to model projective and visibility information as ternary relations. Granularity and complexity of the models can be adjusted: we present two differently detailed realizations and discuss possible applications in Geographic Information Systems.

## 1 Introduction

In the last two decades, qualitative spatial representation and reasoning (QSR) has distinguished itself as a multi–disciplinary, full-fledged research area aiming at understanding mechanisms underlying human spatial cognition (cf. Cohn and Renz 2008, for an overview). The mainstream of this research has focused on the definition and analysis of so-called qualitative spatial calculi: formal theories capable of resembling human spatial representation and reasoning abilities.

P. Fogliaroni (✉)
Department for Geodesy and Geoinformation, Vienna University of Technology, Gusshausstr. 27-29, 1040 Vienna, Austria
e-mail: paolo@geoinfo.tuwien.ac.at

E. Clementini
Department of Industrial and Information Engineering and Economics, University of L'Aquila, via G. Gronchi 18, I-67100 L'Aquila, Italy
e-mail: eliseo.clementini@univaq.it

© Springer International Publishing Switzerland 2015

M. Breunig et al. (eds.), *3D Geoinformation Science*,
Lecture Notes in Geoinformation and Cartography,
DOI 10.1007/978-3-319-12181-9_15

Literature offers many examples of qualitative calculi modeling different aspects of space like topology (Randell et al. 1992), direction (Freksa 1992; Frank 1991; Clementini et al. 2010), and distance (Hernandez et al. 1995). The majority of these theories have been designed to model entities embedded in either 1D or 2D space, like points, line segments, and regions. However, we live in a 3D (4D, considering temporal dimension) universe and, while some aspects can be satisfactorily approximated into a lower-dimensional space, there exist certain spatial properties that are inextricably related to all three spatial dimensions. One example is visibility.

Visibility is defined[1] as "the ability to see or be seen" and is the expression of the probably most important sensory experience through which we cognize space: visual perception. Visibility is a fundamental property in many research and applied fields, spanning from spatial cognition to architectural design, urban planning, and computer graphics.

Visibility is, intrinsically, a three-dimensional concept and to satisfactorily capture its semantics we cannot reduce it to lower-dimensional spaces, unless it is intended as a projection of 3D space on the plane, like in the approach of Tarquini et al. (2007) where visibility is defined on a 2D projective framework[2]. This approach, indeed, only works for overly-simplified environments where certain properties of physical objects can be ignored but that poorly fit real world variety. For example, in a world consisting only of right prisms, visibility can be modeled in a 2D reference system enriched with height information. Simply by allowing our objects to have holes or concavities, such a simplified modeling does not suffice any longer. Thus, capturing semantics of visibility in real-world scenarios calls for full 3D modeling.

This paper introduces a frame of reference for 3D space that allows for empowering, refining, and simplifying previously-presented qualitative models for projective and visibility relations (see Sect. 2 for details). The presented frame of reference lies upon two reference solids and is anchored to a direction vector. The first referent represents an observer that, in most cases, is intended to be a sensing agent (human or robotic). The second referent represents an obstacle, possibly occluding the observer's sight field. The direction vector can be the expression of gravity force: "an important asymmetric factor in the world" (Franklin and Tversky 1990). If the observer assumes a different vertical axis than that of gravity force—e.g. if the observer is flying or is out of the gravity field—the direction vector can be associated to the feet/head axis of human body or an equivalent axis of a robot.

The remainder of the paper is structured as follows. In Sect. 2 we review related work on qualitative visibility and projective relations in 2D and 3D. Section 3 describes generating elements of the frame of reference and introduces two qualitative models for projective and visibility relations, respectively. In Sect. 4 we

---

[1] http://www.merriam-webster.com/dictionary/visibility.

[2] If living in an imaginary 2D space, visibility would follow quite different rules as in the perilous life experience of Flatland characters (Abbott 1884).

discuss how the model can be applied to objects with holes and concavities. Properties and possible applications of the proposed models are discussed in Sect. 5. Section 6 concludes the paper and describes future work.

## 2 Related Work

Projective relations in 2D space have been modeled by Billen and Clementini (2004) as ternary relations of the form $R_{proj}(A, B, C)$ where $A$ is the located (or primary) object with respect to two reference objects $B$ and $C$. If the convex hulls of $B$ and $C$ are disjoint, it is possible to trace four lines tangent to both referents. These lines are used to partition the plane into five *acceptance areas*,[3] as shown in Fig. 1a. To each acceptance area is associated a so-called *single-tile relation* that holds if $A$ falls completely in that area. The case that $A$ spans multiple acceptance areas is modeled through so-called *multi-tile relations*. Usually, multi–tile relations are named after the and-concatenation of the single tile relations associated to the acceptance areas spanned by the primary object. The number of multi-tile relations equals the number of admissible combinations of single-tile relations. Whether a combination is admissible or not depends on the topology of the objects under consideration. For example, considering connected objects, the relation *before* ∧ *after*$(A, B, C)$ is not admissible. Indeed, it is impossible for $A$ to span the areas *Before*$(B, C)$ and *After*$(B, C)$ without intersecting any other area—e.g., *Between*$(B, C)$. The resulting projective model consists of both single– and multi-tile relations whose reasoning properties are discussed in (Clementini et al. 2010).

An extension of the model to cope with solids in 3D space is presented in Billen and Clementini (2006). This work presents two frames of reference grounded on

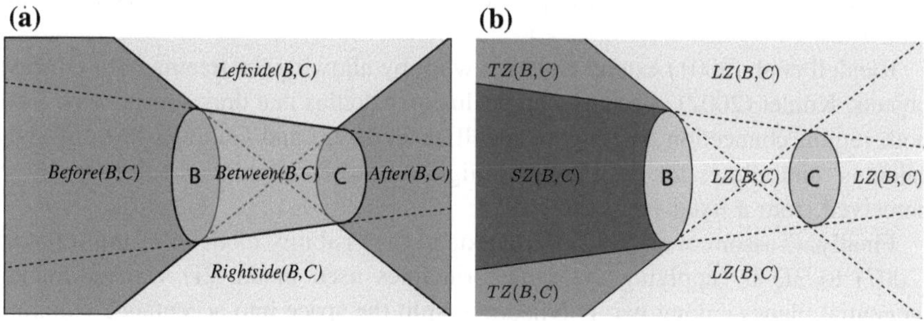


**Fig. 1** Frames of reference for 5-Intersection model Billen and Clementini (2004) (**a**) and visibility model Tarquini et al. (2007) (**b**)

[3] The acceptance area of an *n-ary* relation, can be defined as a parametric subset of the relation's domain, having as a parameter a tuple of $n - 1$ domain objects.

concepts of projective geometry, such as collinearity and coplanarity. The first frame of reference allows for defining four, ternary, single–tile relations: *before*, *between*, *after*, and *aside*. The other frame of reference allows for defining the relative position of a primary object with respect to a configuration of three reference objects and defines the following, quaternary, single–tile relations: *coplanar* (further refined into *internal* and *external*) and *non_coplanar* (further refined into *above* and *below*).

Bartie et al. (2011) integrate the ternary 3D model of Billen and Clementini with other qualitative calculi and use it to represent and reason about the environmental visual perception of an agent modeled as a point. The application of the resulting model to produce cognitively-sound route instructions based on landmarks—as suggested by Raubal and Winter (2002)—is discussed.

A qualitative model for visibility relations in 2D space is introduced in Tarquini et al. (2007). The provided relations express the property of an observed (primary) object of being seen from an observer when a third object acts as an obstacle (reference objects). Inspired by the frame of reference for projective relations presented by Billen and Clementini (2004), the authors design the frame of reference depicted in Fig. 1b. In this frame, the observer object (*C*) is imagined to be a source of light and the acceptance areas are named accordingly: light zone (*LZ*), shadow zone (*SZ*) and twilight zone (*TZ*). The single–tile, ternary, relations associated to these zones are called: *Visible*, *Occluded*, and *PartiallyVisible*, respectively.

Fogliaroni et al. (2009) refine the visibility model by also considering the direction where a partially visible object is perceived with respect to the obstacle. Drawing upon this finer-grained model, they define a qualitative coordinate system based on visibility and show its application for robotic localisation and navigation. De Felice et al. (2010) exploit the model for environmental learning.

A different strategy to modeling visibility is adopted by Galton (1994): The viewpoint of the observer is fixed in space and the arrangements of obstacle and observed object are described by binary relations. In this way Galton defines a lines of sight calculus providing 14 qualitative relations and discusses its reasoning properties.

Randell et al. (2001) extend Galton's work by allowing the treatment of concave objects. Köhler (2002) introduces an occlusion calculus that draws inspiration from both region connection calculus (Randell et al. 1992) and Galton's lines of sight calculus, and allows for describing configurations of two convex solids in 3D as perceived from a fixed viewpoint.

Finally, (Tassoni et al. 2011) try to extend the visibility model of Tarquini et al. (2007) to 3D by applying the same principles used in the 2D variant: tracing tangential planes among two polyhedra to split the space into acceptance volumes. They discover that the approach becomes cumbersome as the number of tangent planes is not fixed (contrarily to linear tangents in 2D) and depends on the shape of the objects as well as on their relative position and orientation.

## 3 A Frame of Reference for Projective Relations in 3D Space

In this section we present a frame of reference generating from two reference solids (*B* and *C*) and a direction vector (**d**) that allows for partitioning 3D space in different manners. In the next two sections we show how two such partitions can be used to model relative directions and visibility, respectively.

We assume, without loss of generality, that the direction vector **d** is parallel to the *z* axis of a canonical Cartesian coordinate system. Moreover, we assume that the convex hulls of the two reference objects (*B* and *C*) are disjoint.

Before proceeding, let us shortly recall some basic concepts of analytic geometry of space. A plane $\pi$ in 3D Euclidean space $\mathbb{E}^3$ is a flat surface that is uniquely identified by one of its points $P_0$ and an orthogonal non-zero vector **n**—cf. Fig. 2a. Thus, a plane is the locus of points $P = (x, y, z)$ such that $(P - P_0) \cdot \mathbf{n} = 0$. Assuming that $P_0 = (x_0, y_0, z_0)$ and that the vector $\mathbf{n} = a\mathbf{i} + b\mathbf{j} + c\mathbf{k}$ is expressed in terms of its components along an orthonormal basis, we have that:

$$\pi : ax + by + cz + d = 0 \tag{1}$$

with $d = -ax_0 - by_0 - cz_0$. Any plane $\pi$ splits $\mathbb{E}^3$ into two parts called *half–spaces* and identified by the following inequalities:

$$\pi^+ : ax + by + cz + d > 0 \tag{2}$$

$$\pi^+ : ax + by + cz + d < 0 \tag{3}$$

where the half–space $\boldsymbol{\pi}^+$ is the one "aimed" by the vector **n**—cf. Fig. 2b.

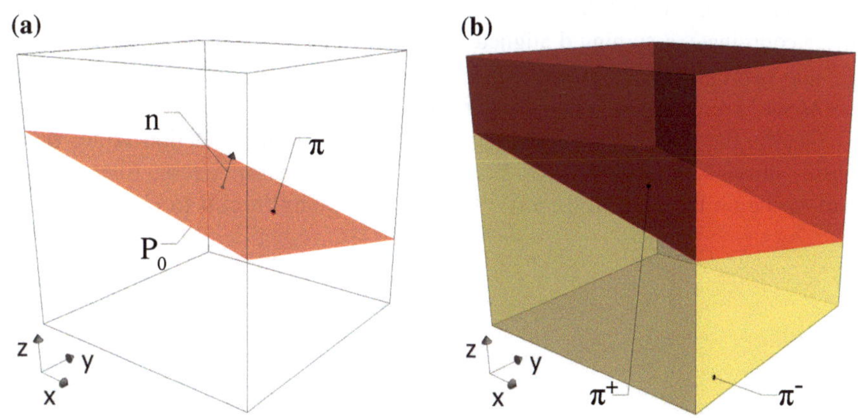


**Fig. 2** A point *P* and a vector **v**, uniquely define the plane $\pi$ passing through *P* and orthogonal to **v**. Point-vector definition of a plane in $\mathbb{E}^3$ (**a**) splits the space into two half–spaces (**b**)

For simplicity, assume that $B$ and $C$ are spheres. Call $\Pi$ the infinite set of planes tangential to both. Assume each such plane $\pi$ is given in a point–vector representation, with the point being the intersection point with $C$ and the vector going from this point to the centre of $C$. That is, $C$ falls completely in the half–space $\pi^+$—except its intersection point with $\pi$. Tangential planes can be split into two disjoint sets: (i) the set of *external tangent planes* $\Pi_{ext}$ consists of all the planes $\pi$ such that the other reference object $B$ also falls into $\pi^+$; (ii) the set of *internal tangent planes* $\Pi_{ext}$ consists of all the planes $\pi$ such that $B$ falls into $\pi^+$.

We now have all the constructive elements to define the frame of reference shown in Fig. 3f consisting of two pairs of cones[4] and three planes.

The first pair of cones $\Delta_{ext}$, shown in Fig. 3a, are called *external cones*. The cone $\Delta^+_{ext}$ (resp. $\Delta^-_{ext}$) is the subspace obtained by intersecting the half–spaces $\pi^+$ (resp. $\pi^-$) identified by all the external tangent planes $\pi \in \Pi_{ext}$:

$$\Delta^+_{ext} = \bigcap_{\pi \in \Pi_{ext}} \pi^+; \quad \Delta^-_{ext} = \bigcap_{\pi \in \Pi_{ext}} \pi^- \tag{4}$$

Note that, when $B$ and $C$ have same radius, the cone $\Delta^+_{ext}$ degenerates into a cylinder and $\Delta^-_{ext}$ does not exist—i.e. is the empty subspace.

Figure 3b depicts the second pair of cones $\Delta_{int}$, called *internal cones*. The cone $\Delta^+_{int}$ (resp. $\Delta^-_{int}$) is the subspace resulting from the intersection of the $\pi^+$ (resp. $\pi^-$) half–spaces defined by all the internal tangent planes $\pi \in \Pi_{int}$:

$$\Delta^+_{int} = \bigcap_{\pi \in \Pi_{ein}} \pi^+; \quad \Delta^-_{int} = \bigcap_{\pi \in \Pi_{int}} \pi^- \tag{5}$$

Figure 3c shows how the two pairs of cones are located with respect to each other.

The plane *Above* (resp. *Below*), cf. Fig. 3d, is the external tangent plane whose intersection point with $B$ has max (resp. min) value along the direction vector **d**—i.e., $z$ coordinate, assuming **d** aligned to $z$ axis. Note that *Above* (resp. *Below*) splits $\mathbb{E}^3$ into two half–spaces: *Above*$^+$ (resp. *Below*$^+$) is the half–space containing $B$ and $C$; *Above*$^-$ (resp. *Below*$^-$) is the other half–space.

Finally, the plane *Central*, cf. Fig. 3e, is parallel to the direction vector **d** and passes through the centroids of $B$ and $C$. We call *Central*$^+$ (resp. *Central*$^-$) the half–space on the right (resp. left) of *Central*, as going from $B$ to $C$.

---

[4] Pyramids, if considering polyhedral objects.

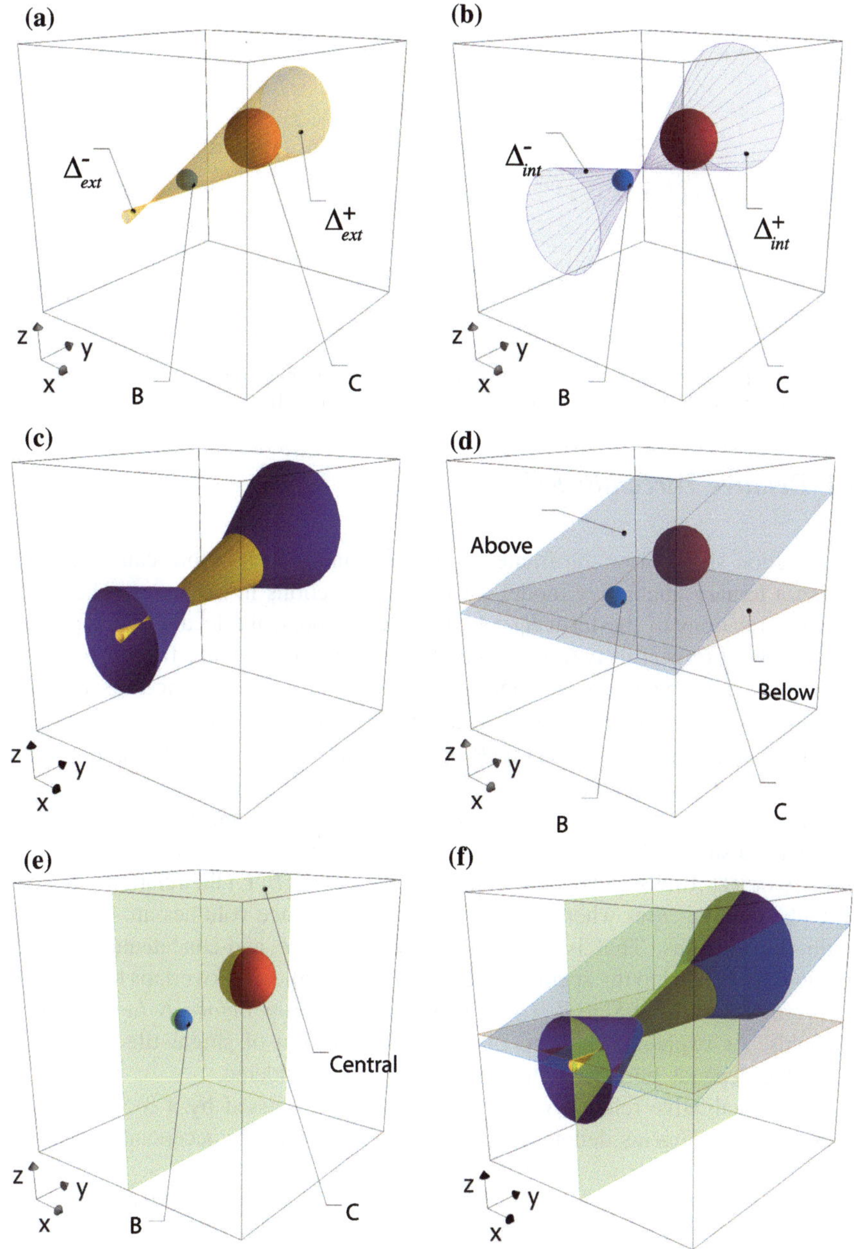


**Fig. 3** Constructing elements of the frame of reference. **a** External cones. **b** Internal cones. **c** External and internal cones. **d** Plane *Above* and *Below*. **e** Plane *Central*. **f** Frame of reference

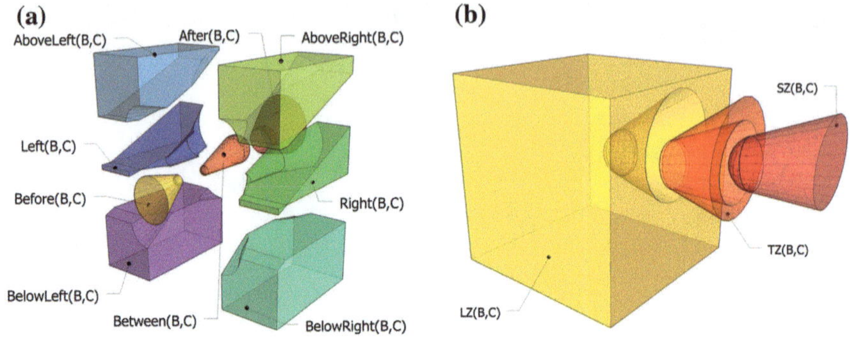


**Fig. 4** Two possible interpretations of the frame of reference of Fig. 3f. **a** Acceptance volumes for relative directions. **b** Acceptance volumes for visibility (exploded)

## *3.1 Relative Direction Relations*

This section shows how the frame of reference introduced before can be used to design a ternary qualitative model for relative directions in 3D space. The model provides relations of the form $R_{dir}(A, B, C)$ that express the location of a primary object $A$ with respect to two reference objects $B$ and $C$, heading from $B$ to $C$.

Besides the constructive components of the frame of reference discussed in section above, we resort to the Convex Hull operator to partition $\mathbb{E}^3$ into nine volumes parametric with respect to $B$ and $C$, as shown in Fig. 4a. The definition of the volumes is given in Table 1.

These volumes are referred to as *acceptance volumes*[5] and to each of them is associated a single-tile direction relation named accordingly. So, for example, if the primary object $A$ falls completely in the volume *Before*($B$, $C$) the relation *Before*($A$, $B$, $C$) holds. The cases when $A$ spans multiple acceptance volumes are covered by multi–tile relations. That is, relations named after the and-concatenation of the relations associated to the spanned volumes. For example, if $A$ overlaps the volumes *Before*($B$, $C$) and *LeftSide*($B$, $C$), the multi–tile relation *Before* ∧ *LeftSide*($A$, $B$, $C$) holds. As explained in Sect. 2, which combination of single–tile relations is allowable depends on the topology of the objects considered.

Every qualitative relation $R_{dir}(A, B, C)$ can be expressed by a 3 × 3 Boolean matrix whose elements denote the intersection of $A$ with the acceptance volumes:

$$\begin{pmatrix} A \cap AboveLeft(B,C) & A \cap After(B,C) & A \cap AboveRight(B,C) \\ A \cap Left(B,C) & A \cap Between(B,C) & A \cap Right(B,C) \\ A \cap BelowLeft(B,C) & A \cap Before(B,C) & A \cap BelowRight(B,C) \end{pmatrix}$$

[5] Acceptance volumes are the generalization to 3D space of the notion of acceptance areas, discussed in Sect. 2.

**Table 1** Definition of the acceptance volumes depicted in Fig. 4a

| Volume | Definition |
|---|---|
| *Between(B, C)* | *ConvexHull(B* ∪ *C)* |
| *Before(B, C)* | $\Delta^-_{int}$*ConvexHull(B* ∪ *C)* |
| *After(B, C)* | $\Delta^+_{int}$*ConvexHull(B* ∪ *C)* |
| *AboveLeft(B, C)* | $(Central^- \cap Above^- \cap Below^+)\backslash(\Delta_{ext} \cup \Delta_{int})$ |
| *Left(B, C)* | $(Central^- \cap Above^+ \cap Below^+)\backslash(\Delta_{ext} \cup \Delta_{int})$ |
| *BelowLeft(B, C)* | $(Central^- \cap Above^+ \cap Below^-)\backslash(\Delta_{ext} \cup \Delta_{int})$ |
| *AboveRight(B, C)* | $(Central^+ \cap Above^- \cap Below^+)\backslash(\Delta_{ext} \cup \Delta_{int})$ |
| *Right(B, C)* | $(Central^+ \cap Above^+ \cap Below^+)\backslash(\Delta_{ext} \cup \Delta_{int})$ |
| *BelowRight(B, C)* | $(Central^+ \cap Above^+ \cap Below^-)\backslash(\Delta_{ext} \cup \Delta_{int})$ |

## 3.2 Visibility Relations

This section presents a different interpretation of the frame of reference presented before that provides a ternary qualitative model for visibility in 3D space. The model provides spatial predicates of form $R_{vis}(A, B, C)$ capturing the semantics of the visibility relation holding between an observer (B) and an observed object (A) when a third, opaque, object (C) acts as an obstacle.

Given two reference objects *B* and *C* it is possible to split $\mathbb{E}^3$ as depicted in Fig. 4b. We follow the naming convention adopted by Tarquini et al. (2007) that, imagining the observer to be a source of light, distinguishes three main volumes (whose definitions are given in Table 2):

> **Light Zone (LZ)** According to the source–of–light metaphor, this is the well–lighted subspace, or, in other terms, is the set of points visible from any point of the observer. That is, it is possible to trace a straight line segment between any point of the observer *B* and any point of LZ(*B*, *C*) such that no point of the obstacle *C* belongs to such a segment. If an object *A* falls into this zone we say it is *visible* from *C* and denote it by *Visible(A, B, C)*.
> **Shadow Zone (SZ)** Conceptually, this can be imagined as the shadow cone casted by the obstacle *C* when the observer *B* is a light source. SZ (*B*, *C*) is the subspace whose points are occluded from every point of the observer *B* by the obstacle *C*. That is, it is impossible to trace a straight line segment between any point of *B* and any point of SZ(*B*, *C*) that does not intersect *C*. If an object *A* falls into this zone we say it is *occluded* from *B* by *C* and denote it by *Occluded(A, B, C)*.

**Table 2** Definition of the acceptance volumes depicted in Fig. 4b

| Name | Volume | Definition |
|---|---|---|
| Light zone | *LZ(B, C)* | $\mathbb{E}^3\backslash(\Delta^+_{int}\backslash(ConvexHull\,(B \cup C)\backslash C))$ |
| Shadow zone | *SZ(B, C)* | $(\Delta^+_{ext}\backslash\Delta^-_{int})\backslash(ConvexHull\,(B \cup C)\backslash C)$ |
| Twilight zone | *TZ(B, C)* | $\Delta^+_{int}\backslash\Delta^+_{ext}$ |

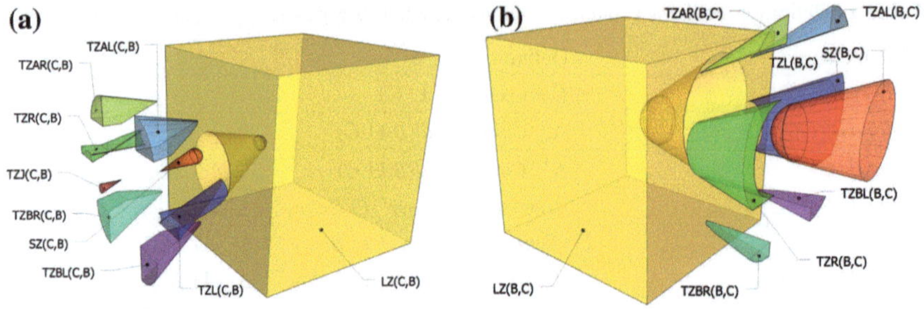


**Fig. 5** Acceptance volumes for visibility relations (exploded). **a** The case of a finite shadow zone (*SZ*) also yields the twilight zone joint (*TZJ*). **b** An infinite shadow zone (*SZ*) and the partition of the twilight zone according to directional primitives

**Twilight Zone** (**TZ**) This is a transition zone between well–lighted and obscured parts of the space. It is the set of points that are occluded from some points of the observer *B* but visible from others. If an object *A* falls into this zone we say it is *partially visible* from *B* and denote it by *PartiallyVisible*(*A*, *B*, *C*).

Besides this coarse partition, a finer-grained one can be obtained from the same constructive elements that splits Twilight Zone into smaller acceptance volumes. This is similar to the refinement in granularity operated on the 2D model in Fogliaroni et al. (2009) and allows for distinguishing where the *partially visible* object is perceived with respect to the obstacle. We use the plane *Central*—cf. Fig. 3e—to distinguish between left and right side and the planes *Above* and *Below* —cf. Fig. 3d—to differentiate among objects perceived above and below the obstacle, respectively. The resulting partition is depicted in Fig. 5b; names and definitions for the refined Twilight Zones are provided in Table 3. The qualitative relations associated to these zones are named *Partially Visible*∗, where ∗ is a placeholder for *AboveRight*, *AboveLeft*, and so on. The semantics of these partially visible relations is straightforward: let us assume, for example, that an object *A* falls into the volume TZL(*B*, *C*)—cf. Fig. 5b. Then, *B* sees *A* partially and perceives it on the left side of *C*: *PartiallyVisibleLeft*(*A*, *B*, *C*). Such a finer distinction turns very helpful for certain applications, as will be discussed in Sect. 5.

It is worth noting that the Shadow Zone can be finite or infinite according to the relative size and position of the reference objects. This is depicted in Fig. 5a, where the roles of observer and obstacle are swapped. In this case two main things are worth to be noted. (i) Shadow Zone does not extend to infinity. (ii) It is possible to distinguish a new zone called Twilight Zone Joint (TZJ): if an object *A* falls into this zone, it is partially visible and perceived "all around" (i.e., simultaneously AL, AR, L, R, BL, and BR) the obstacle *B*. We denote this by *PartiallyVisibleJoint*(*A*, *B*, *C*).

Similarly to the case of relative directions, visibility relations $R_{vis}(A, B, C)$ can be single– and multi–tile and can be represented by a Boolean matrix of void or non–void intersections. For the coarse model we need a 1 × 3 matrix of the kind:

**Table 3** Definition of the refined twilight zones depicted in Fig. 5 in terms of the coarse visibility relations in Table 2 and the of the direction relations in Table 1

| Name | Volume | Definition |
|---|---|---|
| Twilight zone left | *TZL(B, C)* | $TZ(B,C) \cap Left(B,C)$ |
| Twilight zone right | *TZR(B, C)* | $TZ(B,C) \cap Right(B,C)$ |
| Twilight zone above left | *TZAL(B, C)* | $TZ(B,C) \cap AboveLeft(B,C)$ |
| Twilight zone above right | *TZAR(B, C)* | $TZ(B,C) \cap AboveRight(B,C)$ |
| Twilight zone below left | *TZBL(B, C)* | $TZ(B,C) \cap BelowLeft(B,C)$ |
| Twilight zone below right | *TBRR(B, C)* | $TZ(B,C) \cap BelowRight(B,C)$ |
| Twilight zone joint | *TZJ(B, C)* | $\Delta^{-}_{ext}$ |

$$(A \cap LZ(B,C) \quad A \cap TZ(B,C) \quad A \cap SZ(B,C))$$

For the finer–grained model we need a 3 × 3 matrix of the kind:

$$\begin{pmatrix} A \cap TZAL(B,C) & A \cap LZ(B,C) & A \cap TZAR(B,C) \\ A \cap TZL(B,C) & A \cap TZ(B,C) & A \cap TZR(B,C) \\ A \cap TZBL(B,C) & A \cap TZJ(B,C) & A \cap TZBR(B,C) \end{pmatrix}$$

# 4 Treating Holes and Concavities

So far we assumed that the objects to be modeled are convex and hole–free. For the sake of an easier dissertation, we also assumed that they are spheres and that the convex hulls of the reference objects are disjoint. However, in real cases (e.g. in a 3D Geographic Information System) objects are usually modeled as generic solids, possibly with holes and concavities. While the treatment of solids homeomorphic to spheres is no special case, the treatment of solids with holes and/or concavities must be discussed.

First, consider the case that only the primary object has holes or concavities. In this case the frame of reference—thus the acceptance volumes of the two models—is not affected and the only difference concerns the multi–tile relations that can be deemed allowable. For example, if *A* is convex and without holes, the relation *Left* ∧ *AboveLeft* ∧ *AboveRight* ∧ *Right(A, B, C)* can never hold. Indeed, in order to intersect the acceptance volumes corresponding to this multi–tile relation, *A* should also necessarily span either *Before(B, C)*, or *Between(B, C)*, or *After(B, C)*. Yet, this is no longer the case if *A* is concave—c.f. Fig. 4a for a visual aid.

Now, consider the case that reference objects are not convex and hole-free. We first address the case of concave objects and, to tackle it, we suggest to (i) decompose concave objects into a number of convex parts such that $B = B_1 \cup \cdots \cup B_m$ and $C = C_1 \cup \cdots \cup C_n$; (ii) compute the acceptance volumes for each pair *(Bi, Cj)*; (iii) aggregate the obtained volumes by applying different aggregation rules according to the semantics of the relation each volume is associated with. Using this approach allows for reducing the problem of treating concavities to that of treating multi-objects (i.e. objects consisting of several parts). Convex decomposition has been widely treated in the field of computational geometry and optimal methods have been found (Chazelle 1984). Computation of acceptance volumes between pairs of parts is no different than applying the model on convex objects. Thus, what remains to be defined are the aggregation rules.

Due to space limitations we only discuss coarse visibility. Moreover, for simplicity, we resort to a 2D example. Since the frame of reference presented in this paper is an extension to 3D of the projective frames of reference in Fig. 2, we stay assured that the same concepts apply to the corresponding 3D relations.

Say the observer $B$ and the obstacle $C$ are concave, call $B_1, \ldots, B_m$ and $C_1, \ldots, C_n$ their convex parts, and denote by $I$ and $J$ the index sets $\{1, \ldots, m\}$ and $\{1, \ldots, n\}$.

First, fix an arbitrary part $C_j$ of the obstacle and aggregate on $B$. The Light Zone (resp. Shadow Zone) is the subspace whose points are *simultaneously* visible (resp. occluded) from every point of the observer. Since the observer consists of several parts one could rather say: to every point of every part. That is:

$$LZ(B, C_j) = \bigcap_{i \in I} LZ(B_i, C_j); SZ(B, C_j) = \bigcap_{i \in I} SZ(B_i, C_j) \tag{6}$$

Twilight Zone can be obtained by subtracting the union of the other zones from $\mathbb{E}^2$.

Now, fix an arbitrary part $B_i$ of the observer and aggregate on $C$. Since the obstacle consists of multiple parts—that is equivalent to have multiple obstacles—the visibility of any point in space is affected by all the parts at the same time. In order to reflect this property on the acceptance zones, one shall apply the following procedure: (a) trace acceptance zones for every pair ($B_i$, $C_j$); (b) remove all the tangents to $B_i$ and $C_j$ that intersect another $C_k$ between $B_i$ and $C_j$; (c) if a tangent to $B_i$ and $C_j$ intersects another $C_k$ beyond $C_j$, cut the tangent at the intersection with $C_k$ and start considering $C_j$ and $C_k$ as a unique part $C_l = C_j \cup C_k$ (because, from the perspective of $B$, $C_j$ and $C_k$ are perceived as a continuum). For a visual exemplification refer to Fig. 6. Enumerate the parts resulting from the application of the procedure above by the index set $K = 1, \ldots, p$. Then we have:

$$LZ(B_i, C) = \bigcap_{k \in K} LZ(B_i, C_k); \quad SZ(B_i, C) = \bigcup_{k \in K} SZ(B_i, C_k) \tag{7}$$

Again, the Twilight Zone can be obtained as set difference.

Finally let us tackle the treatment of objects with holes. The semantics of direction relations is not affected by the fact that one or both reference objects have holes: e.g., the fact that an object $A$ lies on the left of ($B$, $C$) is not affected by the fact that $B$ or $C$ have holes. Conversely, when dealing with visibility relations, the case when the obstacle $B$ has holes must be treated specially. Again, we suggest to look at the obstacle as consisting of different parts: the real object and its holes.

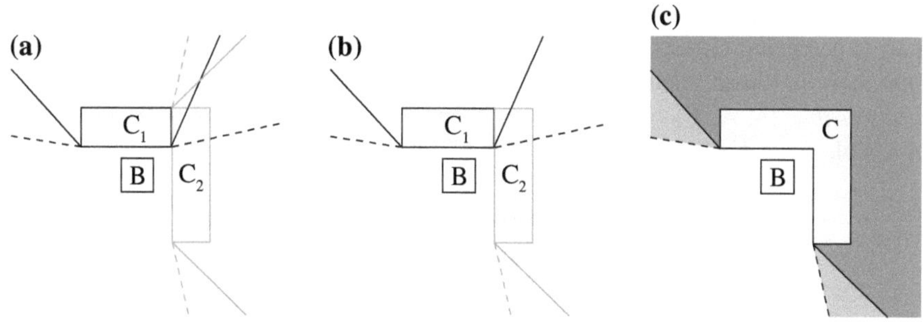


**Fig. 6** The three steps of the aggregation procedure for concave objects

Acceptance volumes must be computed for all parts (which, if concave, must be split into convex subparts), but the role of Shadow and Light Zones must be swapped in the case of holes. Finally, we apply the aggregation rules defined for multi–objects.

# 5 Application to Wayfinding

The models presented in this paper can be applied in a variety of fields, spanning from cognitive robotics to architectural design, urban planning, and computer graphics. In this section we discuss how to generate cognitively-sound route instructions based on landmark visibility, as suggested by Raubal and Winter (2002).

The idea draws inspiration from previous work on qualitative localisation and navigation where Fogliaroni et al. (2009) show that, by repeatedly applying the 2D visibility frame of reference on every pair of objects in the environment, it is possible to split the navigable space into regions. Each such region corresponds to the intersection of a number of visibility acceptance areas, thus, specific visibility relations hold between each point of a region and objects in the environment. In particular, Fogliaroni et al. show that, if the observer is modeled as a point $b$, internal and external tangents collapse into each other, *PartiallyVisible* relations are not defined, and the following properties hold when reasoning on change of perspective:

$$\begin{aligned} Visible(A,b,C) &\Leftrightarrow Visible(b,A,C) \\ Occluded(A,b,C) &\Leftrightarrow Occluded(b,A,C) \\ Visible \wedge Occluded(A,b,C) &\Leftrightarrow PartiallyVisible * (b,A,C) \end{aligned} \tag{8}$$

where ∗ stands for one of the fine–grained partial visibility relations.

Moreover, if *Visible* ∧ *Occluded*(*A*, *b*, *C*), *b* only sees a part of *A* and, in 2D, this part is perceived either on the *Left* or on the *Right* side of *C*. Fogliaroni et al. show that there is a neat relation between these relative directions and the relation *PartiallyVisible**(*b*, *A*, *C*) obtained when changing perspective. Say $A_v$ is the part of *A* visible from *b* then we have the following properties:

$$\begin{aligned} Right(A_v, b, C) &\Leftrightarrow PartiallyVisibleLeft(b, A, C) \\ Left(A_v, b, C) &\Leftrightarrow PartiallyVisibleRight(b, A, C) \end{aligned} \tag{9}$$

Similarly, when operating a change of perspective in 3D, we have that *Right* relations map onto *PartiallyVisibleLeft* ones, *Left* onto *PartiallyVisibleRight*, *Above* onto *PartiallyVisibleBelow*, and *Below* onto *PartiallyVisibleAbove*.

Thanks to these properties, one can say if, how, and in which direction with respect to a certain obstacle, an object is perceived from an observer standing in a visibility region. Fogliaroni et al. also highlight the fact that, since the number of acceptance zones is finite, so is the number of boundaries among them and that for each boundary two qualitative navigation behaviours are defined that lead from one zone to the neighbour and back. Such behaviours can be mapped onto natural language instructions like "walk around *C* keeping it on your right until you start perceiving *A* on *C*'s left", or "walk towards *C*, until *A* disappears from your sight".

So, given a 3D urban dataset *D* consisting of buildings—a subset of which is denoted as landmarks—and of navigable spaces (e.g., roads), one can compute visibility spaces of each landmark *l* by applying the frame of reference of Sect. 3 to every pair (*o*, *l*), where *o* is a building, and intersect them with the walkable spaces.

Using a standard routing algorithm for pedestrian routes one can (i) compute the shortest path *p* from *A* to *B*, (ii) intersect it with the visibility spaces of landmarks, and (iii) generate natural language instructions by considering which boundaries between visibility volumes are crossed by *p* as going from *A* to *B*.

## 6 Conclusions and Future Work

We presented a qualitative 3D frame of reference that is both, an extension of the 2D projective models presented in Billen and Clementini (2004), Tarquini et al. (2007) and a refinement of the 3D models discussed in Billen and Clementini (2006).

Billen and Clementini (2006) presented two qualitative models for projective relations, one ternary and one quaternary. Thanks to the anchorage to a direction vector, the relative direction model presented in Sect. 3.1 allows for maintaining a lower arity (ternary) while augmenting expressiveness (differentiating relations along the vertical direction).

In Sect. 3.2 we showed how the newly introduced frame of reference allows for extending the visibility model presented in Tarquini et al. (2007) and the finer–grained variant discussed in Fogliaroni et al. (2009).

We also discussed how the two models we introduce can be applied when considering objects with holes and concavities and, finally, we speculated on a possible application on landmark-based route instructions.

For future work we plan to investigate how the presented models can be characterised in terms of current standards for 3D modeling in GIS (e.g., CityGML). Moreover, we intend to compute reasoning tables in order to perform qualitative reasoning.

## References


Abbott EA (1884) Flatland: a romance of many dimensions. Dover Publications Inc., New York (republished in 1992)

Bartie P, Reitsma F, Clementini E, Kingham S (2011) Referring expressions in location based services: the case of the 'opposite' relation. In: Advances in conceptual modeling. Recent developments and new directions. LNCS. Springer, pp 231–240

Billen R, Clementini E (2004) A model for ternary projective relations between regions. In: Advances in database technology. LNCS. Springer, pp 537–538

Billen R, Clementini R (2006) Projective relations in a 3D environment. In: Geographic information science. Springer, pp 18–32

Chazelle B (1984) Convex partitions of polyhedra: a lower bound and worst-case optimal algorithm. SIAM J Comput 13(3):488–507

Clementini E, Skiadopoulos S, Billen R, Tarquini F (2010) A reasoning system of ternary projective relations. IEEE Trans Knowl Data Eng 22(2):161–178

Cohn AG, Renz J (2008) Qualitative spatial representation and reasoning. In: Handbook of knowledge representation, foundations of artificial intelligence. Elsevier, pp 551–596

De Felice G, Fogliaroni P, Wallgrün JO (2010) Qualitative reasoning with visibility information for environmental learning. In: Proceedings of the 6th international conference on geographic information science (GIScience 2010)

Fogliaroni P, Wallgrün JO, Clementini E, Tarquini F, Wolter D (2009) A qualitative approach to localization and navigation based on visibility information. In: Proceedings of the 9th international conference on spatial information theory. LNCS. Springer, pp 312–329

Frank AU (1991) Qualitative spatial reasoning with cardinal directions. In: Proceedings of the 7th austrian conference on artificial intelligence. ÖGAI. Springer, pp 157–167

Franklin N, Tversky B (1990) Searching imagined environments. J Exp Psychol Gen 119(1):63

Freksa C (1992) Using orientation information for qualitative spatial reasoning. In: Theories and methods of spatio-temporal reasoning in geographic space, vol 639. LNCS. Springer, pp 162–178

Galton A (1994) Lines of sight. AI Cogn Sci 94:103–113

Hernandez D, Clementini E, Di Felice P (1995) Qualitative distances. In: Spatial information theory a theoretical basis for GIS, vol 988. LNCS. Springer, pp 45–57

Köhler C (2002) The occlusion calculus. Cognitive vision workshop

Randell D, Cui Z, Cohn AG (1992) A spatial logic based on regions and connection. In: Principles of knowledge representation and reasoning: proceedings of the 3rd international conference (KR'92). Morgan Kaufmann, Massachusetts, pp 165–176

Randell D, Witkowski M, Shanahan M (2001) From images to bodies: modelling and exploiting spatial occlusion and motion parallax. In: IJCAI, pp 57–66

Raubal M, Winter S (2002) Enriching way finding instructions with local landmarks. In: Geographic Information Science. LNCS. Springer, pp 243–259
Tarquini F, De Felice G, Fogliaroni P, Clementini E (2007) A qualitative model for visibility relations. In: KI 2007: advances in artificial intelligence, pp 510–513
Tassoni S, Fogliaroni P, Bhatt M, De Felice G (2011) Toward a qualitative model of 3D visibility. 25th international workshop on qualitative reasoning (IJCAI 2011) (position paper)

MIX
Papier aus verantwortungsvollen Quellen
Paper from responsible sources
FSC® C105338

If you have any concerns about our products,
you can contact us on
**ProductSafety@springernature.com**

In case Publisher is established outside the EU,
the EU authorized representative is:
**Springer Nature Customer Service Center GmbH**
**Europaplatz 3, 69115 Heidelberg, Germany**

Printed by Libri Plureos GmbH
in Hamburg, Germany